Governance of Seas and Oceans

From the *Seas and Oceans* Set
coordinated by
André Mariotti and Jean-Charles Pomerol

Governance of Seas and Oceans

Edited by

André Monaco
Patrick Prouzet

WILEY

First published 2015 in Great Britain and the United States by ISTE Ltd and John Wiley & Sons, Inc.

ISTE Ltd
27-37 St George's Road
London SW19 4EU
UK

www.iste.co.uk

John Wiley & Sons, Inc.
111 River Street
Hoboken, NJ 07030
USA

www.wiley.com

Library of Congress Control Number: 2015952413

British Library Cataloguing-in-Publication Data
A CIP record for this book is available from the British Library
ISBN 978-1-84821-780-5

Contents

Chapter 7. Integrated Management of Seas
and coastal areas in the Age of Globalization

Yves HENOCQUE and Bernard KALAORA

Chapter 8. Ocean Industry Leadership
and Collaboration in Sustainable
Development of the Seas

Paul HOLTHUS

Foreword

We have been asked by ISTE to stimulate work in the area of the environment. Therefore, we are proud to present the "Seas and Oceans" set of books, edited by André Monaco and Patrick Prouzet.

Both the content and the organization of this collection have largely been inspired by the reflection, initiatives and prospective works of a wide variety of national, European and international organizations in the field of the environment.

The "oceanographic" community, in France and internationally – which is recognized for the academic quality of the work it produces, and is determined that its research should be founded on a solid effort in the area of training and knowledge dissemination – was quick to respond to our call, and now offers this set of books, compiled under the skilled supervision of the two editing authors.

Within this community, there is a consensus about the need to promote an interdisciplinary "science of systems" – specifically in reference to the Earth's own "system" – in an all-encompassing approach, with the aim of providing answers about the planet's state, the way it works and the threats it faces, before going on to construct scenarios and lay down the elementary foundations needed for long-term, sustainable environment management, and for societies to adapt as required. This approach facilitates the shift of attention from this fundamental science of systems (based on the analysis of the processes at play, and the way in which they interact at all levels and between all the constituent parts making up the global system) to a "public" type of science, which is finalizable and participative, open to decision-makers, managers and all those who are interested in the future of our planet.

In this community, terms such as "vulnerability", "adaptation" and "sustainability" are commonly employed. We speak of various concepts, approaches or technologies, such as the value of ecosystems, heritage, "green" technologies, "blue" chemistry and renewable energies. Another foray into the field of civilian science lies in the adaptation of research to scales which are compatible with the societal, economic and legal issues, from global to regional to local.

All these aspects contribute to an in-depth understanding of the concept of an ecosystemic approach, the aim of which is the sustainable usage of natural resources, without affecting the quality, the structure or the function of the ecosystems involved. This concept is akin to the "socio-ecosystem approach" as defined by the Millennium Assessment (http://millenniumassessment.org).

In this context, where the complexity of natural systems is compounded with the complexity of societies, it has been difficult (if only because of how specialized the experts are in fairly reduced fields) to take into account the whole of the terrestrial system. Hence, in this editorial domain, the works in the "Seas and Oceans" set are limited to fluid envelopes and their interfaces. In that context, "sea" must be understood in the generic sense, as a general definition of bodies of salt water, as an environment. This includes epicontinental seas, semi-enclosed seas, enclosed seas, or coastal lakes, all of which are home to significant biodiversity and are highly susceptible to environmental impacts. "Ocean", on the other hand, denotes the environmental system, which has a crucial impact on the physical and biological operation of the terrestrial system – particularly in terms of climate regulation, but also in terms of the enormous reservoir of resources they constitute, covering 71% of the planet's surface, with a volume of 1,370 million km^3 of water.

This set of books covers all of these areas, examined from various aspects by specialists in the field: biological, physical or chemical function, biodiversity, vulnerability to climatic impacts, various uses, etc. The systemic approach and the emphasis placed on the available resources will guide readers to aspects of value-creation, governance and public policy. The long-term observation techniques used, new techniques and modeling are also taken into account; they are indispensable tools for the understanding of the dynamics and the integral functioning of the systems.

Finally, treatises will be included which are devoted to methodological or technical aspects.

The project thus conceived has been well received by numerous scientists renowned for their expertise. They belong to a wide variety of French national and international organizations, focusing on the environment.

These experts deserve our heartfelt thanks for committing to this effort in terms of putting their knowledge across and making it accessible, thus providing current students with the fundaments of knowledge which will help open the door to the broad range of careers that the area of the environment holds. These books are also addressed to a wider audience, including local or national governors, players in the decision-making authorities, or indeed "ordinary" citizens looking to be informed by the most authoritative sources.

Our warmest thanks go to André Monaco and Patrick Prouzet for their devotion and perseverance in service of the success of this enterprise.

Finally, we must thank the CNRS and Ifremer for the interest they have shown in this collection and for their financial aid, and we are very grateful to the numerous universities and other organizations which, through their researchers and engineers, have made the results of their reflections and activities available to this instructional corpus.

André MARIOTTI
Professor Emeritus at University Pierre and Marie Curie
Honorary Member of the Institut Universitaire de France
France

Jean-Charles POMEROL
Professor Emeritus at University Pierre and Marie Curie
France

Transformations in International Law of the Sea: Governance of the "Space" or "Resources"?

1.1. Introductory remarks

In researching primary legal issues, and the legal instruments promoted by them enabling the governance of seas and oceans, the International Law of the Sea occupies an extremely important place. In both its ancient and current forms, it represents a foundation of rules and solutions utilized by States with coastal borders to impose maritime controls on marine waters. This Law of the Sea has almost wholly determined the current structure of administrative and legal divisions traced on the waters by governments and certain organizations. In this exercise, the concept of "marine spaces", and especially of "marine spaces" to which Law of the Sea is applicable, has been essential. A very large portion of governments' rights to act on the surface and beneath the seas depends on these spaces (section 1.2), and, most often, what is done with resources located in the seas (living or mineral resources) is also a result of them (section 1.3). The link between these two aspects must be explained, as they are increasingly intertwined. It is a transformation that involves considerable concerns regarding marine resources.

Chapter written by Florence GALLETTI.

1.2. The importance of marine spaces in International Law of the sea

It is advantageous for us to define Law of the Sea, which determines the legal governance of seas and oceans, (section 1.2.1). This will help us to show the difference instilled between "marine" zones and "maritime" zones (section 1.2.2) and, whether it is public or private intervention on the seas and oceans that is intended, this slight difference is a fully operational one. The evolution of the Law of the Sea and the usages made of it by governments reveals the ongoing legal hold of coastal States over marine spaces; this is practised in various, rhizomatic forms – that is spread out and sometimes creeping, but in which the distance to the coast (via the legal concept of the "baseline") remains an essential point, and the horizontal division of marine waters both under the jurisdiction of States or beyond it, a strong constant (section 1.2.3).

1.2.1. Definitions of International Law of the sea: a keystone of the governance of maritime spaces

The question of governance of maritime spaces cannot be set without a definition exercise. In a restricted sense, it is a set of institutions, legal rules and processes enabling the adoption of an institutional and legal framework for action, and then the development of related public or private interventions, on the delinated space. Despite its importance, the International Law of the Sea is often poorly defined, or defined by default by differentiating it from other, more sector-specific legal disciplines pertaining to activity at sea. It is related in particular to maritime law, a very ancient concept used in the past to address issues arising both from private laws having to do with maritime activity and international public law for marine activities [PON 97]. This has resulted in widespread (and quite understandable) confusion. Today, however, maritime law pertains mostly to the specific commercial activity of maritime shipping, and is defined as "all legal rules pertaining to navigation on the seas" [ROD 97] or as "all legal rules pertaining to private interests engaged at sea"[1] [SAL 01]. More rarely,

---

1 [SAL 01, p. 389].

some specialists attribute a broader definition to maritime law, seeing it, for example, as "all rules pertaining to the various relationships having to do with the utilization of the sea and the exploitation of its resources² [LØP 82a], or study it in parallel with International Law of the sea³. However, the two subjects are separate. The International Law of the Sea addresses seafaring activities in a more complete manner; these naturally include navigation, but from another angle, which can bring the two types of law together and render them complementary. The International Law of the Sea, widely referred to as such since the first Geneva Conference on the Law of the Sea in 1958, is more relevant to matters of governance of spaces at sea. With it, oceans and seas are not without legal rules and arguments; on the contrary, a field of law is specifically dedicated to them [DAU 03].

One of its definitions presents it as "all rules of International Law pertaining to the determination and subsequent status of maritime spaces, and pertaining to the system of activities framed by the marine environment"⁴ [SAL 01]. A more geopolitically oriented definition presents it as "Law regulating relations between States concerning the utilization of the sea and the exercise of their power over maritime spaces"⁵ [LØP 82b]. Both of these definitions emphasize a *spatial* element that is highly determinative of the holding of rights by governments and of the exercise of these rights in relation to other governments.

The context of the Law of the Sea involves the pre-eminent position of the "State" in several senses. The central government is a favored subject in International Law, alongside the various international organizations in which this quality is recognized⁶ [DAI 02]. Because it is situated under the aegis of general International Law, the Law of the Sea obeys the same operating principles, those of an "international legal order" in which States remain vital actors but are very free for the creation of multilateral or bilateral legal rules. It results from this that the State is the vector of the rules making up a system of governance applied to its continental, applied to its continental or island territory, and to the marine spaces that are extensions of these

---

2 [LØP 82a, p. 77 and s.], cited by Rodière, Pontavice [ROD 97].
3 See the highly exhaustive book by Beurier [BEU 14].
4 [SAL 01, p. 375].
5 [LØP 82b, p. 49] cited by Rodière, Pontavice [ROD 97].
6 Daillet and Pellet refer more extensively on this point [DAI 02].

(adjacent maritime spaces). It is vector directly influenced by International Law or by its own inventiveness and (most often) within the limits of action permissible by written (conventional) or customary International Law. Outside of these marine the vector spaces under State control, concepts such as "right to fly flag and flag law" or recourse to "nationality" are all forms of extension – on the high seas – of the national Law of a State (or an institution such as the European Union (EU)) over often far-flung waters which are no longer linked by geographic proximity and legal bonds "of sovereignty" or "of jurisdiction" between the State and these marine spaces.

1.2.2. *Marine spaces considered by law: the interest of qualifying maritime zones*

All marine spaces, as far as they are able to be distributed, identified and described by life sciences or biogeography, for example, are not all spaces considered by law. The existence of seas and oceans is a fact that can be understood scientifically, but the existence of a Law of the Sea associated with these bodies of water does not necessarily follow from this. For this to occur, a shift is required between the term "marine zones" and the concept of "maritime zones". In geographical terms, a "marine" or "maritime" zone – the terms are used almost interchangeably – may designate any part of the sea of some geographic sector in which a given activity takes place; this means that we see for example that gulfs, coastal areas, and shorelines are designated but without any legal consequence [LUC 03][7]. When the desire or obligation for public intervention and regulation of an area of marine zones arises, legal definition exercises take place.

In legal terms, the concept of a "maritime zone" designates a marine zone or marine space to which a legal system is applicable. The legal term "maritime zone" is applicable only to marine spaces, each corresponding to its own legal system[8] [LUC 03]. Thus, via various successive conventions and conferences on the Law of the Sea, a large number of maritime zones have been established by coastal States according to the legal marine spaces predefined in the conventions, of which the most recent and consequential was the United States Convention on the Law of the Sea (UNCLOS)[9] of

---

7 According to Lucchini [LUC 03, p. 11].
8 According to Lucchini [LUC 03, p. 12].
9 United Nations Conference of the Law of the Sea.

December 10, 1982, sometimes also known as the Montego Bay Convention (MBC). In addition to common maritime zones which have now become relatively classic, such as internal waters[10], territorial seas[11], contiguous zones[12], exclusive economic zone (EEZ)[13], continental shelves[14], high seas[15] and the international zone of seabed called "the Area", there are now maritime zones arising from the first zones and thus from least ambitious rights of establishment according to the legal adage "he who can do more can do less", such as fishing zones, ecological protection zones (EPZs), and possibly integrated management coastal zones (IMCZs) [GHE 13], etc. To all this, we must also add specific configurations of marine spaces which the Law of the Sea has sanctioned and to which it has granted, subject to compliance with certain conditions, a legal status that gives rise to specific legal effects: islands[16], bays[17], straits[18], international canals, low-tide elevations[19], archipelagic waters[20], etc. (such as in the Philippines or Indonesia; see Figure 1.1). The definition of these marine spaces is not only a simple typology conveniently available for coastal States wishing to have them recognized or established for their own benefit; but, it is always accompanied by a legal system of rights and obligations regarding maritime zone x for the State concerned (coastal State, port State, flag-holding State, with adjacents coasts, etc.) [PAN 97]. These situations can be more complex; a double legal system can exist in one maritime space, with the typical case being that of territorial waters (or two adjoining territorial seas) containing a strait used for international navigation, such as the Strait of Bonifacio between France and Italy. If the analysis of spaces greatly affects the delimitation of fishing activity or navigation (two activities that are particularly highly developed and sanctioned in the Law of the Sea [LUC 90, LUC 96b]), the question of marine resources, their protection and their development also plays a role.

---

10 Art. 8 CNUDM.
11 Art. 2 and 4 CNUDM.
12 Art. 33 CNUDM.
13 Part V of the CNUDM, art. 54-75.
14 Art. 76 to 85 CNUDM.
15 Part VII CNUDM.
16 Art. 121 CNUDM.
17 Art. 10 CNUDM.
18 Part III, CNUDM.
19 Art. 13 CNUDM.
20 Art. 46-49 CNUDM.

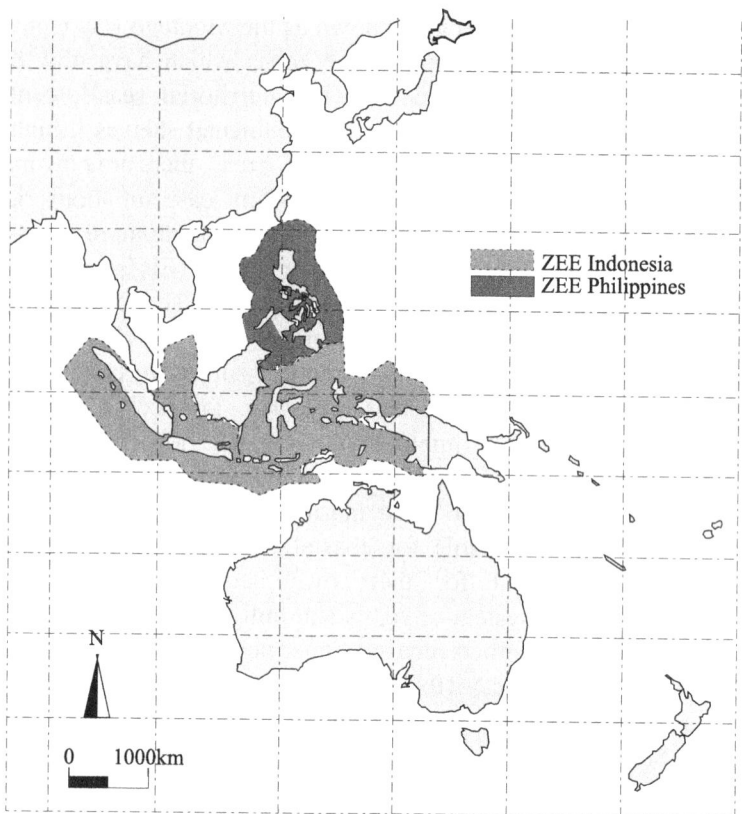

Figure 1.1. *Archipelagic waters and exterior limits of the two EEZs of two archipelagic States in the sense of International Law of the Sea (Indonesia and the Philippines, 2013) (source: www.vliz.be, adapted from Thema Map software, 2012, https://themamap.greyc.fr) (document does not presuppose any support for the claims of governments), from [GAL 15]*

1.2.3. *Development of legal control over certain marine spaces: a phenomenon both ancient and renewed*

The Law of the Sea is a very ancient consideration, and a perennial discipline marked with key historic points. This historic link between the sea as a route of transport and the securitization of commercial activities was already present in the Roman period and is contained in the expression *Mare nostrum*; the end of the 15th Century saw intercessions centered on the sharing of the oceans (the 1494 Treaty of Tordesillas between Spain and Portugal, typically with an Atlantic partition), and spatial oppositions between protagonists concerning access and use of the seas; first in the 16th Century with Spanish authors, and the burgeoning 17th Century has remained notorious for its famously controversial proclamation by James I, King of England, prohibiting access to the North Sea for foreign vessels (a recurring problem in English seas), which was greeted by two opposing doctrines on the possible appropriation of sea spaces and the applicability of prohibitions of this type, Hugo De Groot's "Mare Liberum" in 1609 and John Selden's "Mare Clausum" in 1635. Though it did not prevent control over areas quite distant from the coasts (for example, the 18th Century Hovering Acts in England), the principle of freedom of the seas has been triumphant in relative terms (all States were given the minimum right to navigate and trade, as described in Philip Meadows's 1689 treatise) since the late 17th Century and remains in effect even today, as it is applied to modern activities conducted by countries and their nationals on the seas (the six freedoms of the high seas).

The 20th Century was characterized by the affirmation of the sovereignty of States over spaces and natural resources located further and further away from the coasts, a trend first seen in matters of customs, or what we would qualify as customs today (for example, the Liquor Treaties of the United States in the early 20th Century), and then more generally beginning in 1937, and clearly used by States after 1945. In the United States, President Truman's proclamation on American policy concerning the resources of the soil and subsoil of the continental shelf and in territorial waters (known as the Truman Proclamation and dated September 28, 1945) represented a public declaration of the maritime control that national governments could have, express and exercise [APO 81]. This was taken up and furthered by regionalist expansionist doctrines, so to speak, including those of several South American States, beginning in 1947 and continuing today. With decolonization, marine space, with its exploitable resources and consequent ability to guarantee the economic development of new States, has become a strategic concern for both developing and developed countries [GAL 11]. The latter are witnessing a reduction in maritime zones not under the jurisdiction of a government, and consequently must both rethink legal relationships controlling access to these spaces that have now been taken over by others, and step up their own controls over marine spaces situated in such a way as to be extensions of their land territory. The view, however, inexact in a legal sense, that maritime expansion is simply an extension of maritime territories as a prolongation of a state's sovereignty over its continental land holdings [QUE 97] has been used to justify tendencies toward ever-widening control. This, for water columns, involves an outside limit of a State's EEZ that has now reached 200 NM[21] from the baseline and an of a State's EEZ outside limit of the continental shelf also set at 200 NM for general cases, barring (in a generalized manner) a request for extension of the continental shelf to 350 NM or even slightly more, in the event that certain geomorphological characteristics are present [TAS 13].

The appearance and development of interest in marine spaces beyond areas of national jurisdiction seem to be characteristic of the 21st Century so far; or perhaps it is more correct to say that the current century has reawakened them [DEM 09, MAR 14], particularly via questions regarding the effectiveness of collective governance measures undertaken for rezoning in maritime zones on the high seas for specific purposes (for example, fishery areas and the competence of institutions associated with this zoning and this sector of activity overall), or having to do with the opportunity for the evolution of the Law of the sea in order to enable the future creation of new maritime zones within the high seas (zoning for the purposes of environmental protection). Yet, this focus on marine spaces beyond jurisdiction zones originated in the 1970s, with the initiative introduced by Arvid Pardo in the United States to include on the agenda for the 22nd session of the UN General Assembly, the question of the peaceful use of seabeds and their exploitation outside jurisdiction zones (August 17, 1967). This was followed by a number of transformations: the creation of the "International seabed zone" called the Area, mandate of the International Seabed Agency[22], responsible for regulating this zone (the ISA is headquartered in Jamaica) and the legal system governing these seabeds and activities of exploration and later of exploitation that went along with it. These changes are sometimes later criticized by authors and practitioners of law of exploitation of the sea because they are fairly remote from the philosophy of the conservation, protection and development of common heritage of humankind, which was upheld at the start but of which little remains today. However, they are all part of this heritage, in which the consideration of spatial elements has taken priority of place to the detriment of other factors.

---

21 One marine mile = 1,852 m = one nautical mile = 6,076 feet. Here, M. is used as an abbreviation for the marine mile used in marine maps. The abbreviation Nq is also used for nautical miles. French-language books on the Law of the Sea usually use the abbreviation M.M. (marine mile) and English-language books use N.M. (nautical mile).
22 ISA – International Seabed Authority.

1.2.4. *Maritime zones near and far from coasts: a distinction established between systems of sovereignty and those of jurisdiction*

1.2.4.1. *Origins*

The impossibility of establishing a single legal system for the oceans has led to a fragmentation of spaces. This situation, described both above and below, is in part the product of so-called "customary" International Law, but above all of the "conventional" International Law of the Sea. The conventional or written source, with the increase in international conventions and in the numbers of signatories to them, has supplanted the traditional source: in 2014, there were 166 States or organizations that had ratified or were adhering to the UNCLOS, for example. It remains the case that some States, and not the lesser ones in terms of their maritime capacity, still function for the most part under customary International Law (for example, the United States). The two sources of law have converged as a result of the effort made by written International Law to codify a number of practices and translate them into written provisions, and of efforts made in practice to comply with or move closer to the written provisions, which are becoming increasingly universal, pertaining to maritime zones, maritime delimitations, etc.

The process of codifying International Law was first undertaken in 1924 and continued by the Hague Conference in 1930. Subsequent benchmark events are well known; in the domain of the Law of the Sea and fishing, they occurred in 1958, 1960, 1973, 1982, 1994, etc., dates which correspond to the 1st United States Conference on the Law of the Sea, held from February 24 to April 27, 1958 in Geneva, and to the four associated international conventions signed on April 29, 1958: the 1958 Geneva Convention on Territorial Sea and Contiguous Zone (CTS)[23], the April 29, 1958 Geneva Conference on Fishing and the Conservation of Living Resources on the High Seas (CFCLR)[24], the 1958 Geneva Convention on the High Seas (CHS)[25] and the 1958 Geneva Convention on the Continental Shelf (CCS)[26]. Subsequent dates correspond to the 2nd United States Conference on the Law

---

23 Entered into force on September 10, 1964.
24 Entered into force on March 20, 1966.
25 Entered into force on September 30, 1962.
26 Entered into force on June 10, 1964.

of the Sea, held from March 16 to April 26, 1960, and to the 3rd United Nations Conference on the Law of the Sea, the highly exhaustive work of which, lasting from 1973 to 1982, resulted after 9 years of exchanges between States in the United States Convention on the Law of the Sea of December 10, 1982 (UNCLOS), which did not become effective until November 16, 1994. This period from 1973 to 1982 corresponded to a rewriting of the Law of the Sea into a monumental text: the "Constitution of Oceans" (followed by related agreements). This shaped what has since usually been referred to as the "new Law of the Sea" [QUE 94].

1.2.4.2. *Confirmation*

This "new Law of the Sea", which has been approved by a growing number of the world's States, includes legal marine spaces [VIN 08] that have been rendered more uniform:

– concerning first coastal zones in the broad sense; these include "internal waters" and then "territorial sea" with a current maximum breadth of 12 NM, or 22.2 km, under the sovereign governance of a State. Sovereignty rights are attached to these two maritime zones and are recognized as belonging to coastal States; they include a wide range of powers allocated to governmental bodies competent in the maritime domain;

– possibly followed by the "contiguous zone", the span of which toward the sea must not exceed 24 NM from the baseline[27], and, very commonly, the EEZ, the span of which toward the sea must not exceed 200 NM from the baseline (an EEZ must have a span – in the direction of the open sea – of 200 NM that is less than or equal to 370 km drawn from the baseline). These are the so-called waters "under jurisdiction", subject to the recognized jurisdiction rights of coastal States. Fishing zones of x NM, ecological protection zones of x NM or zones of various appellations of x NM are thus incorporated into waters under jurisdiction, provided that they are situated outside the exterior limit of territorial waters and within a distance of less than 200 NM toward the open sea, measured from the baseline (Figure 1.2, in white). Here the challenges for coastal States in establishing and causing to be recognized a baseline[28] as far as possible from the coastline become

---

27 In the hypothetical event that territorial waters of 12 NM. remain 12 NM. maximum of open sea for a contiguous zone.
28 Baselines are addressed in the United Nations Convention on the Law of the sea (UNCLOS) in articles 5, 7, 14, 47, etc.

understandable, as this means so much maritime mileage gained in the direction of the open sea when the baseline diverges from the coastline;

– next comes the "high seas". This zone, in the hypothetical event of maximum maritime control exercised by a coastal States, begins after the exterior limit of the EEZ, at more than 370 km from the baseline. However, in the hypothetical event of maritime control reduced to simple territorial waters with no other zone established by the States as an extension, the high seas may begin immediately at the outside limit of the territorial waters, thus beginning very near the coast; distances between the baseline and the start of the high seas can thus be variable depending on the configuration of maritime coasts and the expansionist desires of States;

– the "(legal) continental shelf"[29], which is a separate configuration from the water column, can be considered a legal marine space. It has been progressively acknowledge that this can be recognized for up to 200 NM, thus generating sovereignty rights for the States that holds it – but only up to this maximum of 200 NM. It is of little importance that the geomorphological continental shelf extends beyond these 200 NM. In reality, the legal continental shelf begins after the outside limit of a territorial sea/territorial waters, which goes back to the statement that the soil and subsoil of territorial waters, while forming the start of a geomorphological continental shelf, are not tied to the legal reasoning of the International Law of the Sea with regard to the legal continental shelf. This does not affect their fate because, since the soil and subsoil of territorial seas are in territorial waters, the State exercises incontestable sovereignty rights over them. Their legal system of internal law varies according to States[30]. After territorial waters, the next part of the geomorphological shelf begins to be considered as the legal continental shelf, which initiates the application of the legal system of the continental shelf and the States's sovereignty rights over this shelf. In the end, there is, therefore, no break in the treatment of this geomorphological continental shelf of between 0 and 200 NM in span, because a system of sovereignty rights is applicable, from the start to the outside legal limit of this shelf, but the same fundamental legal principles are not used.

---

29 The adjective is almost always omitted, but it is important for avoiding confusion with the geomorphological shelf.

30 In France, for example, the soil and subsoil of territorial seas constitute elements of the maritime public domain and are covered by the Law of the maritime public domain, while the marine waters of territorial seas do not form part of that domain.

Figure 1.2. *View of territorial waters and EEZ (white) forming waters under sovereignty and under jurisdiction, as opposed to zones outside jurisdiction (light gray) (source: www.vliz.be, adapted from Thema Map software, 2012, https://themamap.greyc.fr). Document does not presuppose any support for the claims of governments, from [GAL 12]*

A rarer situation is the one allowed by the new international Law of the Sea in which a state, or several states jointly, may request to extend its (or their) legal continental shelf to the outer edge of the continental margin (a geomorphological concept); that is up to 350 NM (= 648,200 km) measured from the baseline, or by 100 additional NM (= 185,200 km) calculated from the 2,500 m isobath linking all points situated at 2,500 m of depth. The system applied is still the one of sovereignty over the legal continental shelf. In the event of agreement granted by the Commission on the Limits of the Continental Shelf (CLCS) to several States following their joint request, all that remains for these States is to mark out among themselves the lateral portions of the shelf belonging to each of them.

1.2.4.3. *Principles of more uniform outlines but with varying configurations*

From the previous considerations, there results a marking-out of maritime spaces that is never exactly the same, if only because of the geography of coastlines and the skill needed to trace the baseline and to have this outline accepted by other States. It is chosen by the State, which remains free, but within the limits of the legal possibilities offered by current international Law of the Sea – as well as, importantly, the (geopolitical) risks arising from overly ambitious maritime claims. With regard to the definition of limits of territorial waters, States have generally extended the limit of their territorial seas from 3 NM or 6 NM to 12 NM, not hesitating to take advantage of the possibilities offered by 20th Century Law of the Sea. The end of the old system of narrow territorial seas was planned, but exceptions remain; Greece has a territorial sea of only 6 NM, as does Turkey. With regard to the contiguous zone, almost 75 States have claimed an extension of 24 NM measured from the baseline, with this zoning composed of 12 NM of territorial waters and 12 NM of the contiguous zone. Other States have proven less ambitious, such as Venezuela, whose territorial waters and contiguous zone do not exceed 15 NM in total, while others have taken greater spans (for the contiguous zone) or represent specific cases (notably, North Korea, with its military strip of 50 NM). As for the marking-out of the high seas, this always begins at a minimum distance from the coastline which varies depending on whether a state does or does not desire maritime zones outside its territorial waters. Finally, certain differences are due to the geographical and legal constraints provided for by the Law of the Sea itself; this is the case for semi-enclosed seas [GAL 15a], distinguished from large oceanic spaces in the UNCLOS text (articles 122 and 123). In article 122, semi-enclosed seas are those "surrounded by several states and linked to another sea or to an ocean by a narrow passage, or composed entirely or in part of territorial waters and the zones of economic exclusivity of multiple States" (Figure 1.3).

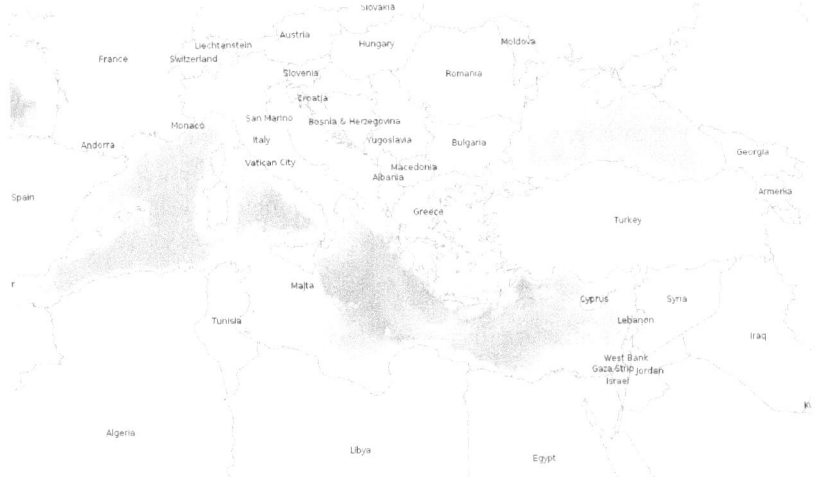

Figure 1.3. *Semi-enclosed seas in the sense of article 122 of the UNCLOS (source: Méditerranée et mer Noire, adapted from Demis NL software, 2014, www.demis.nl). Document does not presuppose any support for the claims of governments*

In comparison to oceanic spaces adjoining the coasts of states that can claim them, here the particular characteristics of semi-enclosed seas have led to the consideration of legal systems better suited to the exercise of the competences of coastal States. Unilateral action on the sea by bordering states was allowed with increasingly frequency throughout the 20th Century. This has been combined with the idea of shared seas (which is not the sharing of seas). Sharing is not synonymous with appropriation that excludes use by others. In international texts, the idea of sharing has been maintained as a way of ensuring the freedom of a maximum number of users to develop activities. Today, sharing often means joint responsibility for deteriorations and for the instruments to be mobilized, two points underlying the International collaboration required from states and the ways in which they are required to participate in collective forms of marine resource management. Thus, cooperation between States is explicitly recommended by article 123: they "must cooperate with one another in the exercise of the rights and the execution of obligations belonging to them under the terms of the Convention". In this context, bordering States and those with adjacent coasts have often limited themselves with regard to control, due to lack of space and in order not to relinquish the smallest share of space on the high seas. This attitude is in the process of changing, for example in the western Mediterranean, with the recent EEZ declared in 2012 by France and in 2013 by Spain [GAL 12], which have created significant legal problems (with regard to both the plotting of outlines and to rights) and are undoubtedly harbingers of an acceleration of this phenomenon, and the possibility of the disappearance of the high-seas maritime zone in the Mediterranean [ROS 12a]. This would be a revolution in the history of the theoretical conception and practice of the Law of the Sea; in the meantime, what is happening is a rebalancing, for the benefit of States bordering semi-enclosed seas, spatial situations inherited from the 3rd United States Conference (1973–1982) and encouraged by it, which marked "the triumph of the oceanic State" [LUC 84].

This approach of the Law of the Sea using maritime space and zoning is vital. It has been so historically (as it has provided an opportunity for numerous full point developments), pacifically (as it goes back to the origins of tension among States and has contributed to the resolution of disputes between States[31]), and above all in relation to the more environmental forms

---

31 Resolutions unremittingly pursued by the Law of the Sea under the aegis of the International Court of Justice (ICJ), the International Tribunal on the Law of the Sea (ITLOS), courts of arbitration and "temporary arrangements" between States.

of governance of activities at sea and to the consideration that will be given in future to marine ecosystems and marine resources.

1.3. Place accorded to resources located at sea in the International Law of the Sea

The question of natural resources is a difficult one to address in itself, somewhat like environmental law, the core of which is relatively easy to define but in which the difficulties begin when the outlines must be pinned down. This has to do with the variety of sea resources; the initial opposition in the Law of the Sea between mineral resources and living or biological resources constitutes the fundamental dichotomy (section 1.3.1). The challenges posed by the increasing scarcity of resources are conducive to detailing them. The analysis also becomes one of the intertwining of resources, even though they appear to be of the same nature. One trend in the analysis and evolution of law is to separate fishery resources from other living resources, or to differentiate – or even set against one another – targeted fishery resources from non-targeted species, or bycatch, resulting from an initial fishing operation (single-species or multi-species) conducted as part of a legally defined fishing activity (deep-water fishing, tuna halieutic, etc.). Catch from non-targeted species or bycatch may make up products derived from fishery products. Finally, the analysis is one of (supposed) ease of exploitation, with differentiations according to simple biological resources (for access, consumption or development) or complex ones such as genetic resources, and for all resources necessitating highly specialized techniques (for example, fishing in extremely deep water and techniques for exploration and the exploitation of non-living resources).

1.3.1. Separate treatment for non-living marine resources and fished living marine resources

One of the fundamental principles of the Law of the Sea is that it ensures the contribution to economic development of states bordering marine spaces and holding marine and coastal resources susceptible to appropriation. It is this principle that has legitimized the Law of the Sea – legal discipline – and which explains the fact that it was massively followed in the 20th Century. The productivist nature of this discipline of law is highly marked, as it enables multiple expansions, such as those of strategic EEZ, to control the legal fate applied to pelagic and benthic resources in the water column and on the continental shelf. The advantage of this EEZ lies in the extension of territorial waters; the extent of the rights conferred on a coastal state by an EEZ is clear. The origin of the modern EEZ must not be forgotten. It began in the 1970s, spurred by two sources of pressure – one, the claim by seven Latin American States in favor of an exclusive exploitation zone of 200 NM (the Montevideo Declaration of 1970); and two, Kenya's claim in 1971 before the United States of an EEZ, which marked the first time this new zone was referred to as such.

By declaring an EEZ, a State obtains for itself rights of sovereignty and jurisdiction, but not for the same interventions. Rights of sovereignty are acquired for the management of biological and non-biological resources (conservation art. 61 UNCLOS and details of exploitation art. 62 UNCLOS are relevant) in a water mass and on the bottom of the seabed and subsoil of the sea and for activities of exploration and exploitation (including economic exploitation), based on the currents and tides in the declared EEZ. Rights of jurisdiction are acquired in order to build artificial islands (criminal and civil jurisdictions); to set up or position establishments for fishing or energy production; to enable scientific research; to protect the environment and to establish security zones. These rights are not only civil in nature, but also criminal when they are caused to be respected, and often administrative in matters of authorizing access to and use of the sea floor and subfloor.

Because of this, and without focusing on the EEZ alone, because the legal continental shelf also represents a source of development well understood by States, it has become usual to consider mineral and biological resources separately. They are not at all of the same nature, and the questions they evoke are strongly opposed (even though they are now often grouped together due to the environmental impact inflicted by the exploitation of one on the other). In addition, since the early 2000s, the search for legal and operational manifestations of sustainable development has separated them even further given that the prescriptions for sustainability for non-renewable and renewable natural resources are laid out very differently in International Law and the internal/national law of individual State.

1.3.1.1. *Consideration of certain living marine resources*

The UNCLOS includes a lengthy part XII devoted to biological renewable resources, entitled "Protection and preservation of the marine environment" (art. 192 to 237, UNCLOS), which has influenced the situation of these resources to an extent. However, this part XII is oriented mostly toward questions of multiform pollution (art. 194, 195, 196, 199, 204 to 234 virtually) rather than toward biological resources themselves. One section, "Part VII – Section II – Conservation and management of biological resources on the high seas" is dedicated to these biological resources, and its provisions cover various maritime zones, including the high-seas zone. To the "triumph of the coastal nationalism" [ROS 14][32] it sanctions, UNCLOS also emphasizes the responsibility of states in the management and future fate of marine biological resources. The rights and obligations of individual states in this management are hammered home, and collective and cooperative ways of managing certain marine resources are specified, above and beyond the actions of one State alone.

An example of this may be seen in the text below, which is connected to the UNCLOS and considered an applicative text of it: "Agreement for the purposes of the application of the provisions of the United States Convention on the Law of the Sea of December 10, 1982 relating to the conservation and management of straddling fish stocks and highly migratory fish stocks[33]", from August 4, 1995 [MOM 95] and with an effective date of December 11, 2001. As of 2014, this accord had received 82 ratifications or cases of adherence.

In it, the legal obligation is mentioned to cooperate internationally or regionally by means of commissions and management organizations for certain halieutic activities and marine spaces located partially or wholly

---

32 [ROS 14, p. 871].

33 *The Agreement for the Implementation of the Provisions of the United Nations Convention on the Law of the Sea of December 10, 1982 relating to the Conservation and Management of Straddling Fish Stocks and Highly Migratory Fish Stocks.* The title of this accord is often shortened to the "New York Accord of 1995 on straddling and highly migratory fish stocks"; the category of "highly migratory" fish includes: white albacore tuna (*Thunnus Alalunga*), red tuna (*Thunnus thynnus*), bigeye fatty tuna (*Bigeye tuna or patudo*), stripe-belly bonito (*Listao*), yellowfin tuna (*Thunnus albacares*), black tuna (*Thunnus afianticus*), skipjack tuna (two species), bluefin tuna (*Thunnus maccoyii*), frigate tuna (two species), sea bream (gray dorade), marlin (nine species), sailfish (two species), swordfish, saury, or balao (four species), coryphene or tropical dorade (two species), sharks (six species) and cetaceans (whales and porpoises: six species).

outside zones of water under national jurisdiction (art. 197, UNCLOS). This is both a remit and a request:

– directed toward institutions; all competent international organizations are concerned, on both the regional and global levels, mandated in the areas of fishing, marine environmental protection or even navigation and maritime security; regional fisheries management organizations (RFMOs) are naturally wholly concerned, whatever the spatial jurisdiction (extent) of their competences or the number of species for which they are responsible (single-species or multi-species competence, or for all species) and depending on the strength of the competences they hold, whether they cover one or more marine spaces;

– directed toward conventions and agreements, whether they have access to institutions for the application of its provisions or not. This call for contribution to the application of the rules of UNCLOS to the seas and oceans, including by means of other conventions dedicated to the marine environment, fishing activities or maritime law, shows the superiority of UNCLOS over other legal instruments, which should be understood as complementary to it. Thus, conventions and institutions (instituted before or after UNCLOS) must be in accordance with its spirit, a requirement that is not without difficulties in terms of consistency and cohabitation [IND 13], but it is also equivalent to a sort of general delegation of application, giving the impression that the new Law of the Sea between 1982 and recent years has minimized its involvement in the marine environmental governance of seas and oceans. A significant reawakening on this subject is in progress, with a sort of academic and practical rediscovery of the environmental potentialities of the UNCLOS text (320 articles) and the texts that have flowed it [AND 12, CAS 12].

1.3.1.2. *Consideration of mineral marine resources and the international seabed*

With regard to non-renewable resources, UNCLOS includes a long part XI entitled "The Zone" (art. 13 to 191, UNCLOS). The "Zone", always written with a capital, is here an abbreviation for the International Seabottom Zone. Part XI begins with a definition (art. 133), according to which (1) "resources" are given to mean all *in situ* solid, liquid or gaseous mineral resources which, in the Zone, are found on seabeds or in their subsoil, including polymetallic nodules; (2) resources once extracted from the Zone are called "minerals".

For countries, including some developing ones, the challenges of negotiations to create such a Zone legally and access it were double. On the one hand, it was necessary to allow access to mineral resources (ores, polymetallic modules, cobaltiferous encrustations on underwater mountains and polymetallic sulfurs in volcanic areas, and oceanic ridges marked by hydrothermal processes) and their reservation; and, on the other hand, access to living resources, such as organisms located on hydrothermal sources and in deep-sea trenches with implications for genetic engineering. Beginning in 1967, the fundamental tenets of a legal system for the Zone and the exploitation of seabeds were established.

The meaning of a Zone such as this, initially created to reduce imbalances in conditions between states, support the less endowed and redistribute the wealth, has developed over time. This system, first based on the concept of the common heritage of mankind and the prevention of the appropriation of mineral resources by individual States, has been transformed. This has occurred through revised provisions (the Agreement of July 28, 1994, which prioritizes Part XI of UNCLOS); there are many authors who view this July 28, 1994 Agreement as a loss for developing countries of advances to their benefit, which promised them negotiations and disappeared from the final text. The adoption of this Agreement was accompanied by compromises making it possible to gather the number of signatories necessary for a text to become effective. Since 1994, the status of spaces *beneath* the high seas, situated beyond the 200 NM mark, has been considered in tandem with that of the Zone, with the International Seabed Authority (ISA) supervising and permitting prospectiving activities for the future extraction of mineral, solid, liquid and gaseous resources. The controls provided by the Authority and the financial and technical constraints influencing the filing of requests have not prevented a competitive race to access and share these resources; this involves few requesting parties, but prospecting contracts have been signed since 2000 and their number is growing (in the Clarion-Clipperton fracture board zone, for example, as well as the Indian Ocean). The "Enterprise", a mechanism of the Authority, is permitted to operate on behalf of developing countries and Least Developped Countries (LDC), but except for these cases, the possibilities for LDC remain highly theoretical, since they are always difficult in terms of access to technological transfer, or simply given the current cost of submitting a case for examination by the ISA (approximately $500,000 in 2014).

The most urgent question concerns compatibility between activities exploiting mineral resources and the protection of the marine environment. If we look more closely at this issue, it becomes one involving the way in which the Authority can and will ensure that compatibility measures are taken regarding activities involving ores and the protection of the fragile and little-understood marine environment by operators and requesting parties. This assumes that this compatibility, mentioned in article 145 of UNCLOS, is even possible, which is in no way certain when exploration/extraction and conservation of living organisms must be organized on the same site. Compatibility with other activities carried out in the marine environment (art. 147 UNCLOS) (maritime traffic, etc.) is another form of the question, though a less difficult one. The fact that the ISA has a direct mandate only for mineral resources and not for living resources directly is a complication, and the lack of legal status of marine biodiversity as a whole, are also real pitfall.

This lack does not affect only the field of the exploitation of mineral resources and its immediate and localized environmental consequences. The whole issue of the protection of marine biodiversity is burdened by this lack of legal regulations; above all, it is the portion of these activities qualified as fishery resources that is currently bound by, and its extractions regulated by, fishing laws. Moreover, only a very small fraction of marine flora and fauna species are listed and protected under environmental law on the protection of species.

1.3.2. *Biological resources at the heart of the overlap between environmental law, biological diversity law, the Law of the Sea and fishing law*

Since the Convention on Biological Diversity (CBD) came into effect, biological or living resources have been understood from very specific modern legal points of view (section 1.3.2.1). These do not sit in great harmony with an aging Law of the Sea, though the environments and the management of activities belonging to it go progressively back to the center of the concerns of the modern Law of the Sea. The latter is indeed inevitable, given the dependence of marine resources on the spatial element, and the fact that their legal fate is increasingly determined by it, and the highly spatial character of the Law of the Sea, which remains an unavoidable component of the issue (section 1.3.2.2). This general legal context is also valid for a specific type of resources, fishery resources, and for the fishing activity it has overseen for many years (section 1.3.2.3).

1.3.2.1. *A biological or living resource treated from very specific perspectives*

It is customary to turn to classic instruments of International Environmental Law protecting regional zones, environments or species when a question of biological diversity arises in general, and when one concerning faunistic and floristic resources arises in particular. Successive environmental conventions and agreements have generated obligations and motivations (soft law) for the rational and sustainable management of living resources utilized by States and their citizens. The number of conventions signed and ratified bears witness to States' willingness to submit to a legalized organization of access to resources and to the legal processes applied to it. For the management of species, and later for their protection, some texts are quite old, such as the International Whaling Convention of December 2, 1946, or highly mediatized, such as the Convention on international trade in endangered species of wild flora and fauna of March 3, 1973 (CITES[34]) and the Bonn Convention on migratory species of June 23, 1979[35] and the agreements resulting from it[36]. Incontestably, the Convention on biological diversity, adopted following the Rio Conference in June 1992 and made effective very rapidly on December 29, 1993, has modified perspectives and relationships with regard to biological diversity. It shares the designation of universal convention with UNCLOS and boasts more signatory States than the latter (167 Parties for UNCLOS and 193 for CITES). Each of these has given rise to true progress with regard to the definition of terms vital to the standardization of public interventions in the space; with regard to the elements composing biological diversity[37]; and with regard to the new questions that they pose. With the CBD, and especially the events that followed it (conferences among Parties to the CDB), the conservation of biological resources is becoming a global objective shared

---

34 175 States: it includes three appendices: appendix 1: species threatened with extinction or the effects of commerce; appendix 2: species that may become endangered due to commerce; appendix 3: species declared by one of the parties to be subject to regulation in order to prevent or reduce exploitation.

35 110 States.

36 Example: ASCOBANS Cetaceans: Agreement on the protection of small cetaceans in the Baltic and North Seas (March 17, 1992); ACCOBAMS Cetaceans: Agreement on the protection of cetaceans in the Black and Mediterranean Seas and the adjacent Atlantic zone (1996).

37 Biological diversity: "variability of living organisms of all origins, including terrestrial, marine, and aquatic ecosystems, and the ecological complexes of which they are a part; this includes diversity within species and between species as well as that of ecosystems", art. 2 CDB.

by all countries that hold, supply or explore for biological resources. The CBD, which became effective before the UNCLOS, initially refrained from extending its competence to areas outside the jurisdiction of individual States except in specific hypothetical cases [PRO 07], but in recent years it has continuously placed this question at the center of legal forums and advances to be made. The CBD pursues three principal objectives: the conservation of biological diversity at the national and international levels in order to halt the decline of biological diversity as a whole; the sustainable utilization of its elements; and the fair and equitable sharing of the benefits resulting from the utilization of biological resources, particularly in genetics. Since 2010, multiple subobjectives have been introduced during conferences among parties to the CBD, but they are simply offshoots of these three initial objectives.

In reality, the text of the CBD, on which so many expectations – often very general ones – rest, is intended for the very precise and complex organization of a set of incentives and then legal obligations around access to biological resources and around compensation for this access. To achieve this, the CBD relies on two tools: contract Law with its reciprocal obligations for both parties, which is well suited for redesign; and intellectual property rights (IPRs). The first tool has not yet been perfected, and remains limited in practice, while the second tool has an uncertain future due to the significant arguments opposing it based on the ability to patent a living thing, the protection of inventors' rights, the fair remuneration of provider states, etc. – bearing in mind that IPR, as well as all legal treatment of biodiversity, falls under the multilateral system of trade and commerce organized by the World Trade Organization (WTO), and must come up against it. Moreover, the prevention of conducting trade and the assurance of the most fluid traffic possible of natural resources are impossible except in quite exceptional cases requiring extensive verification before the Dispute Settlement Body (DSB) of the WTO in the event of conflict between States involving these preventions and barriers to the exportation/importation of products. Sea products are subject, like others, to these requirements of International Economic Law.

1.3.2.2. *Recurrence of the dependence of marine "resources" on the spatial element*

How can the fact be explained that the legal consideration of natural resources remains so dependent on the space in which it is found or that holds it? The legal discipline composed of various subjects (Law of the sea, coastal Law, environmental Law, public economic Law, etc.) and their subdivisions (such as the very recent biological diversity Law incorporated into environmental Law, of which it is only a small part, and International fishing Law, which is part of the larger international Law of the sea, etc.) has chosen not to address biological diversity head-on, or in all its aspects. The reaction is not uniform. Law responds to the diversity of biological resources composing it with a variety of legal systems applicable to one type of biological resources and not another. In general, major types of biological resources (for example, "marine" biological resources as opposed to forest biological resources), provided that they have been identified by law (existence of a text defining the "biological" resource, whether it is fishery, genetic, etc.), are distinguished by the possible utilizations that may be made of them: marine resources become "fishery resources" because they are extracted as part of fishing activities for commercialization; other marine resources become "genetic" because they are researched and utilized for pharmaceutical purposes, etc.

At the outset, the principle – taken from International Law – of "the sovereignty of the state over its natural resources" is applicable to natural biological and mineral resources (the state has rights of collection, use, management, destruction, commercialization and control of activities conducted around the biological resource). It possesses them without having to claim them, and it is the state which, via its own internal law, decides whether or not to organize and allow private appropriation (establishment of a system of public or private ownership of these resources on national territory; conditions of compensation in the event of damage to these resources, etc.). However, on the basis of this principle of the sovereignty of a state over its natural resources, which is mentioned in article 193 of UNCLOS, a variety of legal systems have been dedicated to certain resources, leaving others in total or relative legal escheat.

We have also previously seen how the space-distance element at sea determines the legal system of zones and volumes. Faced with marine biological resources in a maritime territory that is divided into several dimensions, the important and recurring question becomes: what is the location of the marine resource and what is the catch location of this marine resource? The fact that the marine resource, or the conflict involving it, is located in territorial waters, a zone of economic exclusivity, on the high seas, or elsewhere, changes both reasoning and treatment. Based on this, marine fishery resources will be subject to a legal system taken from national legislation in matters of fishing (for waters under the sovereignty and jurisdiction of that state) and the prescriptions of International Law in this domain – not to mention, for those belonging to an institution such as the EU, the fishing laws of the EU concerning communal waters belonging to the Member States, or even concerning non-communal waters in which it is desired that European nationals comply with the legal standards set by the EU (high-level standards, or those presented as such). The case of regional fishing organizations (ORGP) participates in the same idea, if we consider the competences they have been granted over functional marine spaces (the perimeters of their competences and control), and the application of the measures they decree when needed. On this basis, it is still advisable to distinguish among the fishery resources listed; those that are more pelagic; those that are highly mobile; and the place where they are located and fished, or the sites they traverse (highly migratory stocks and straddling stocks, which were distinguished by the accord of August 4, 1995, which is as much a part of fishing law as it is of the new Law of the Sea, of which it is an application). The resources listed contrast with forgotten resources, as some of them may be forgotten or ignored by the Laws in effect. If these developments fully overlap with the Law of the sea, reasoning goes well beyond it. For example, taking the case of the collection of genetic resources referred to as *ex situ* (outside of their environment), the preliminary question continues to be: what is the place of origin of the resource, and when did collection occur? Legal treatments will thus vary depending on whether collection took place before or after December 29, 1993, the effective date of the CBD, which introduced new rules.

1.3.2.3. *Fishery resources and fishing rights*

Apart from certain species of marine mammals (art. 65 UNCLOS), resources consisting of so-called anadromous species[38] (art. 66), catadromous species[39] (art. 67), fish stocks found in the EEZ of multiple

---

38 Species that reproduce in freshwaters and migrate toward waters or pass through waters located on the outer limits of the exclusive economic zone of a State other than the country of origin, such as salmon.

39 Species with sea reproduction that spend most of their life in freshwaters of a coastal nation and migrate through the zone of economic exclusivity of another country (eels, for example).

coastal States or simultaneously in the EEZ and in an area adjacent to the zone (straddling stocks[40]) (art. 63), highly migratory species (art. 64), and sedentary species of the continental shelf are expressly mentioned (art. 68). The rest of the species are included in references to biological resources. There are thus none of the additional specifications used by naturalists to describe benthic[41], demersal[42] and pelagic[43] [CUR 11] resources and which form, with numerous categories, marine biological diversity.

Fishing zones falling under the Law of the Sea and Fishing range from rivers to inland waters, and from territorial waters to the high seas and the International seabed Zone in the case of resources on oceanic ridges, hydrothermal hydrothermal vents, etc.

Only some of the trends in the international regulation of marine fisheries [BEE 06, ORE 99, VIG 00] will be discussed here. We will neither touch on spaces located in territorial seas, nor those centered on coasts. However, because large numbers of states have established EEZ (around 100 States possess a 200 NM EEZ) in order to ensure fishery exploitation directly or indirectly (exploitation by others according to systems of fishing agreements and licenses, etc.), it has become usual to address and debate mainly spaces formed by waters under jurisdiction and the legal problems pertaining to them. Moreover, it is the law of high-seas fishing (the law in effect *beyond* waters under national jurisdiction) that has evolved the most, and has been discussed greatly in recent publications and debates.

The legal high seas are still characterized by freedom of fishing on the high seas; this is an ancient principle that persists and recognizes the equality of fishing States in terms of both rights and duties, and the equality of flag states and waterside/coastal states. However, it is a principle that has now been reduced, first by the shrinking of the legal "high seas" area, which covers 64% of the ocean surface, in comparison to the 36% covered by "waters under jurisdiction" (see Figure 1.2 for a visual representation of these surfaces). It also cites a freedom that must be questioned. This freedom

---

40 Halibut, cod and tuna, for example.

41 Species living on the seabed and feeding from the substrate.

42 Species living near the seabed and not far from the coasts and which feed from the seabed or near it (for example, hake).

43 Species living in the surface water layer of oceans (including sardines, anchovies, plankton, etc.).

on the high seas was declared at the same time as its principle was proclaimed, as clearly shown by the texts:

– if we look at the Geneva Conference of April 29, 1958 on fishing and the conservation of biological resources on the high seas, article 1 states that: "All states have the right for their nationals to fish on the high seas", but expressly taking into account "the interests and rights of coastal states" and "provisions concerning the conservation of biological resources";

– if we look at article 116 of the UNCLOS of December 10, 1982, the right to fish on the high seas is given subject to the rights, obligations and interests of coastal states in matters of "marine mammals", particular species ("anadromous" and "catadromous" fish), "straddling fish stocks" and "highly migratory species", and prevalence is given to RFMOs. States are thus not so free anymore, and its nationals are not either, in the face of the historic development of RFMO. This development is a measurable fact, and regional organizations are able to engage in three principal forms of intervention: scientific research; the creation of regulations and measures for fishing; and the possible power of proclaiming fishing bans [BEE 06].

At the same time, articles 117 and following of UNCLOS recall the obligations placed on fishing states in order to ensure the best conservation of species being fished. Some countries interpret these articles "flexibly"; hence the reappearance of the coastal state and its special interests, including on the high seas adjacent to its EEZ, and going so far as the subtle encroaching of the EEZ on the high seas, as the country establishes a presence there to protect the said resources; these attitudes are qualified as creeping or reasoned jurisdiction according to the "presence at sea" doctrine, and are used by some states[44], including when an RFMO holds competence in the high-seas zone concerned. Some States carry out police operations on the high seas in the International fishing organization zone, such as Canada, for species of straddling demersal fish in the zone belonging to the Northwest Atlantic Fisheries Organization (NAFO). However, fishers and boats may come from the flag State, and thus be submitted to its law. Here again, we may note the increased responsibilities placed on the flag State (both on the fisher State, or on the State allowing fishing to fishermens or

---

44 Chile for the common and associated species present in the EEZ of Chile; Argentina for straddling demersal fish and species in the food chain of the species of the EEZ; Russia for the isolated high seas – the Sea of Okhotsk Peanut Hole – in particular a moratorium on yellow spaces, etc.

fishing vessels of other nationality, etc.). The fishing police (officers and Fisheries and Ocean policies enforcement), operating under a flag State, may operate on waters near the flag State's land territory, as well as in waters very far from it (Distant Water Fishing Nationals – DWFNs).

International fishing law has developed significantly since the 1990s, particularly through the impetus of the Food and Agriculture Organization of the United States (FAO). Multiple legal acts, in the form of International conventions, treaties and agreements have been concluded for certain parts of the world, targeting various forms of fishing activity. After the 1984 FAO World Conference on fisheries management and development in Rome, the last decade of the 20th Century was characterized by the search for agreements or instruments of a new type, attempting to generate centrifugal force rather than recognizing additional national desires. This was an effort to bring together the interventions of governing bodies in a marine space that was less high traffic in terms of usages and legislations, and thus the choice fell fatally on marine spaces legally characterized as "high seas". The current challenge is to continue to regulate the treatment there of natural resources moving through these spaces (migratory or straddling fish stocks), or living in them (species localized around sites in the high seas or in the international seabed zone). International initiatives expressed in the legal texts in effect have been a primary argument in favor of acting for states, and a goad to act more quickly. It is in the interest of all of these parties to refer first and foremost to the text of the 1999 FAO code of conduct for responsible fishing, approved by a 4/95 resolution of the October 31, 1995 FAO Conference and with an effective date of December 11, 2001, or to the New York agreement on straddling and highly migratory fish stocks, of the November 24, 1993 FAO agreement of November 24, 1993, effective as of April 24, 2003, aimed at promoting the compliance by high-seas fishing vessels with International conservation and management measures, or to International directives on targeted fisheries and the fight against bycatch of 2010, those on the management of deep-sea fishing on the high seas adopted on August 29, 2008 following a series of FAO technical consultation, and the voluntary ones for the conduct of flag states of February 8, 2013, etc.

These FAO directives, which often appear following advances in industrial technology or new fishing techniques, or following difficulties of definition, such as for deep-water fishing [BES 12], or for illegal, unreported, unregulated (IUU) fishing, and the best knowledge of the effects and scope of these activities, for example, represents a reference suggesting to States, or to regional fishing management organizations, the formulation, enaction and implementation of management measures, such as: protection of species or habitats; consideration of secondary harvesting; information on risks; data collection processes; use of management instruments as in the development of other fisheries; and propositions for regulatory or legislative measures for management alongside traditional technical, engineering or even economic measures applied to the activity. These directives are not mandatory in legal terms, but compliance with them is recommended and encouraged.

The question of unauthorized fishing, which has been referred to as illegal, unreported, unregulated fishery (IUU) fishing[45], was addressed as early as the 1970s, mostly in EEZ. The term "IUU" was first mentioned officially for the first time during the meeting of the Convention on the Conservation of Antarctic Marine Living Resources (CCAMLR) in 1997, and then addressed in the UN Secretary General's report on the oceans and the Law of the sea in 1999, followed by the adoption by the UN General Assembly of a 1999 54/32 resolution including references to the fight against IUU fishing. This was followed by the FAO international plan of action aimed at preventing, forestalling and eliminating IUU fishing (IPA-IUU) of 2001. Three FAO plans on other fishery subjects preceded the 2001 plan (plan for management of fishing capacity, for sharks and for accidental captures of birds by long-line fishing boats between 1999 and 2001). As noted by Leroy [LER 14], this IUU fishing goes back to three different forms of fishing activity involving biological resources, which is important to define clearly [LER 14, ROS 12b]. First, illegal fishing: carried out by national or foreign vessels in waters placed under the jurisdiction of a given state without the authorization of that state, or in contravention of its laws and regulations (for example, fishing equipment, net size, area fished, species, etc.); or by vessels flying the flags of States that are part of a competent RFMO, but which are in infringement of conservation and management measures adopted by this organization, or of national laws or international obligations, including those contractually agreed to by States simply cooperating with a competent RFMO. Next, unreported fishing, which refers to activities that have not been declared to the national authority or competent regional organization, or which have been conducted in a deceptive manner. Finally, unregulated fishing, which includes fishing activities carried out by

---

45 Illegal, unreported, unregulated fishery.

vessels with no national affiliation or flying the flag of a country that is not part of the regional organization responsible for the fishing zone or species in question.

In the debate over the difficulties affecting RFMO, analyses are not systematically interchangeable or reproducible; single-species RFMOs, for example, are not equivalent to tuna-fishing RFMO[46], or to RFMO controlling a wide range of fishery resources. Tuna-fishing organizations have been extensively studied and evaluated in comparison with others, due to the economic importance of tuna-fishing industries. However, the difficulties analyzed and the solutions discussed in the context of their activities remain proper to them.

1.3.3. *Indirect treatment of resources through ecosystem quality conservation policies*

The taking into account of the protection of marine environments is not as recent as the United States media campaigns of the 21st Century would have us believe[47]. Before specific works oriented toward marine biological diversity and the ecosystemic conditions of its maintenance, or toward environmental governance, committed to by the UN Secretary-General, UN institutions, regional commissions of the United States Environment Program, their associated institutional partners, governing bodies and their administrators, and nature protection institutions, it was – internationally

---

46 Tuna fisheries: Indian Ocean Tuna Commission (IOTC); West Indian Ocean Tuna Fisheries Organization (WIOTO); Commission for the conservation of Southern Bluefin Tuna (CCSBT); Commission for the Conservation and Management of Highly Migratory Fish Stocks in the Western and central Pacific (WCPOF), etc. Specific species: North Atlantic Salmon Conservation Organization (NASCO); Pacific Commission on Salmon (PCS); North Pacific Anadromous Fish Commission (NPAFC); International Whaling Commission (IWC); etc. Multiple species: Sub-Regional Fisheries Commission (SRFC), an intergovernmental organization created on March 29, 1985 by convention, with seven member states: Cape Verde, Gambia, Guinea, Guinea-Bissau, Mauritania, Senegal, Sierra Leone; Regional Fisheries Committee for the Gulf of Guinea (COREP); Southeast Atlantic Fisheries Organization (SEAFO); the International Council for the Exploration of the Sea (ICES) (fish stocks in the northeast Atlantic particularly); Center-East Committee for Atlantic Fishing (CECAF); Latin American Organization for the Development of Fishing (OLDEPESCA), etc.
47 Annual reports of the UN Secretary-General "Oceans and the Law of the Sea", "Seas and Oceans" resolution of the UN AG, "Global Conference on the Oceans", "2012 Pact for the Oceans", "Ad Hoc Open-ended Informal Working Group to study issues relating to the conservation and sustainable use of marine biological diversity beyond areas of national jurisdiction".

speaking – the United States Conference on the Environment and
Development (UNCED) in June 1992 that resulted in the Rio declaration on
the environment and development, Agenda 21 – Chapter 17, Oceans – Seas
and Coastlines – 1992. More precisely, from a historical point of view, the
UNEP program for regional seas has played this role since 1974. This legal
system of mobilization for regional seas (18 currently) has given rise – to
name only three–to the Abidjan Convention on cooperation for the
protection and development of the marine environment and "West and
Central African" coastal zones (1981); the Nairobi Convention for the States
of East Africa and the Indian Ocean[48] (1985); and the Barcelona system for
the Mediterranean Sea, developed through the Convention on the protection of
the   marine   environment   and   the   Mediterranean   coastline[49]
and its seven protocols[50], for example. The case of the Mediterranean is
considered to be a very successful one, as the high-seas environment in this
regional sea can be legally protected, due to the Protocol relative to specially
protected areas and biological diversity that has been in effect since late 1999.
A SPA/BD zone may include portions of waters under jurisdiction and
portions of the high seas. This system of regional seas is associated with
partner conventions of the UNEP program for regional seas (the Convention
for the Protection of the Marine Environment in the Baltic Sea Area,
HELCOM[51], and its five attachments; the Convention for the Protection of the
Marine Environment of the Northeast Atlantic, OSPAR (Oslo-Paris) of 1992,
and its five attachments[52], the Bonn Agreement[53] for cooperation in the fight
against pollution by hydrocarbons and other dangerous substances in the
North Sea), and independent conventions: the Convention on the
Conservation of Marine Flora and Fauna of the Antarctic (CCFFMA) of
May 20, 1980 for an oceanic space, and the Convention for the Protection of

---

48 South Africa, Comoros, Reunion, Kenya, Madagascar, Mauritius, Mozambique, Seychelles,
Somalia and Tanzania. Followed by other acts, the Arusha Resolution on the integrated
management of coastal and island areas in eastern Africa (1993), Mahé Declaration (Seychelles)
(1996), etc.

49 February 16, 1976, revised on June 10, 1995.

50 Including the most recent integrated management protocol for coastal zones in the
Mediterranean Sea, effective in 2012.

51 1974 and 1992, bringing together 10 coastal nations of the European Community, which
became the European Union.

52 OSPAR Appendix 1: telluric, OSPAR Appendix 2: pollution prevention via incineration
and immersion, OSPAR Appendix 3: polluting activities offshore; OSPAR Appendix 4:
marine ecological assessment; OSPAR Appendix 5: protection and conservation of BD and
restoration of marine zones.

53 1979, 1983.

the Environment of the Caspian Sea of 2003, bringing together the five States bordering that sea, which is also an inland sea (the Law of the Sea is not applicable to inland seas). If we add to this the action plans of commissions and other international or regional institutions, and the major integrated national marine strategies of States, as well as the development and concretization of European maritime policy, it is necessary to address the indirect treatment of resources, which is done through policies aimed at preserving the quality of ecosystems, which rely on specific and operational zoning with appellations different from those of the classic spaces of the Law of the sea, but incorporate or accommodate them. Among these are:

– maritime zones created under the aegis of the EU, with operational spaces to apply the legal instruments of European-derived Law (European directives and regulations): the space of the four marine subregions of the EU for the application of the Marine Strategy Framework Directive (MSFD) *Directive stratégique cadre pour le Milieu marin* (DSCMM) of 2008, for example; the redivision of zoning related to the definition of coastal waters and marine waters of the member states of the EU; the common waters space, etc.;

– maritime zones created under the influence of international environmental law and maritime security; protected marine areas (PMAs) under various legal appellations and statutes; Ramsar zones from the RAMSAR Convention of 1971 on wetlands of international importance; particularly protected sensitive areas (PPSAs)[54] of the International Maritime Organization (IMO) since 2005; vulnerable marine ecosystems (VMEs) of the FAO, etc.;

– quite recently, but not yet constitutive of functional spaces: the recent ecologically or biologically significant areas/zones (EBSAs), resulting from the 11th meeting of the Conference of Parties (COP) 11 of CBD at Hyderabad in October 2012.

One explanation for the multiplication of these new zoned areas may be technical; they would be more conducive for experimentation with the sectorial policies requiring them.

One of the main ecological objections raised is the overly static character of the protection areas established for coastal or benthic environments and the relatively deskbound resources allocated to them [GAL 14], it being understood that these zones are placed within waters or continental shelves under sovereign

---

54 OMI, Resolution A.982 (24) Revised guidelines for the identification and designation of Particularly Sensitive Sea Areas of December 1, 2005.

governance or within waters under jurisdiction. For pelagic environments, a series of measures of fishing management, limited to a sector of activity, a technique or an area of application, and always limited in time, exists within many zones intended to support various fisheries, it being understood that, here again, they are placed between the coast and the inner limit of an EEZ, and that they are combined (or sometimes absorbed) in PMAs. However, some types of measures have also been instituted in zones on the high seas in the form of fisheries restricted areas, but they do not benefit from the legal status granted to a PMA in the strict sense of the term. The acceptability of restrictions for States other than those which have agreed to the establishment of these areas of restriction continues to be a highly problematic issue, and weakens these efforts at conservation. Using the term "pelagic PMA" to qualify these areas would, therefore, be incorrect. Some authors argue, however, that they would be forms of PMA, since they contribute to some part of fisheries policing [CAZ 12].

It remains the case that, in order to be necessary and even vital, these changes of scale, for example, from individual PMA to networks of PMA, or from microlocalized protections to the legal protection of vast marine ecological networks that are ecologically connected, composed of marine biological corridors and key habitats for species, distributed throughout seas and coasts, are facing difficulties related to classic law of the sea to usual and classical Law of the Sea and financial constraints for public environmental action [GAL 14, GAL 15].

Developments have been in discussion by the United States since 2004 among authors, with the hope of moving forward with these questions of protection beyond zones under jurisdiction [DEM 09], as well as the details of changes to be made to the Law of the Sea: amendments to UNCLOS; new texts to be applied in the form of an agreement (for example, one which would make it possible to establish marine areas on the high seas); regional experimentations with legal acts that would precede the reform of general International Law, etc. More specifically, between January 20 and 23, 2015, the third meeting of the "Ad Hoc Open-ended Informal Working Group" which is a working group of the United Nations was held to study issues relating to the conservation and sustainable use of marine biological diversity beyond areas of national jurisdiction of States. Here, it was decided[55] to

55 Recommendations of the Ad Hoc Open-ended Informal Working Group to study issues relating to the conservation and sustainable use of marine biological diversity beyond areas of national jurisdiction to the 69th session of the General Assembly January 23, 2015.

develop a legally restrictive international instrument that would make it possible to act on these areas beyond national jurisdiction (ABNJ[56]). A preparatory committee, charged with making the principal recommendations to the United States General Assembly (UNGA) on the text project, will begin work in 2016 to complete its suggestions by the end of 2017 and the seventy-second session[57] of the UNGA. These recommendations by the preparatory committee will address four themes identified in 2011: marine genetic resources, including those having to do with the sharing of benefits resulting from their exploitation; instruments for the management of ABNJ zones, including PMAs on the high seas; assessments of the impact on the environment of the high seas; and the strengthening of capacities, including the transfer of marine technology. The seventy-second session of the General Assembly may then summon an intergovernmental conference under the auspices of the United States in order to propose an International legal accord. This procedure is not considered to constitute a questioning of pre-existing legal instruments, or of current global, regional and sectorial frameworks, including some described in this chapter. This is why there is no question here of revolution; rather, it is a matter of very considerable progress in terms of principles. In the end, it is the states that are party to UNCLOS; states that are not signatories but have an interest in this question; members of specialized agencies and certain observers, and any resulting accords or arrangements that will reveal the extent of the true possibilities of such a text.

1.4. Conclusion

The Law of the Sea attempts to provide solutions that are "preventive in order to avoid the emergence of a conflict, and curative if a conflict does occur, in order to resolve this conflict temporarily or definitively" [GAL 11]. It has always prioritized spaces and controls, though it has dedicated itself a great deal to fishing activities. In an increasingly strained geopolitical atmosphere regarding natural resources, the risk of conflict has recurred in a permanent manner, or at least that seems to be the case. For example, it is likely that climatic changes are rendering the high seas less favorable for the fishing of tuna species, which have been widely trawled, and that tuna resources have been moving differently since the redistribution of the EEZs

---

56 Areas beyond national jurisdiction, or ANBJ.
57 2014 marked the 69th session of the General Assembly of the United Nations.

of states and on the high seas known before now (the Indian Ocean may be particularly affected according to various scenarios), which threatens halieutic profits, changes the exploitation and control capacities of the countries concerned and of regional fishing organizations; challenges legal and economic arrangements between coastal States affected; and will require a modification of the forms of fishing agreements agreed upon between states and foreign fleets [GAL 15]. It will be advisable to know how to change from control to management, to use the words of Professor Lucchini [LUC 82], and to the management of activities impacting marine natural resources, if it proves impossible to manage marine species freely – a self-evident fact that is often forgotten.

1.5. Bibliography

[AND 12] ANDREONE G., CALIGIURI A., CATALDI G., *Droit de la mer et émergences environnementales*, Editoriale Scientifica, Naples, 2012.

[APO 81] APOLLIS G., *L'emprise de l'Etat côtier*, Pedone, Paris, 1981.

[BEE 06] BEER-GABEL J., "La réglementation Internationale des pêches maritimes", *Oceanis*, Institut océanographique de Monaco, vol. 32-1, pp. 78–110, 2006.

[BES 12] BESLIER S., "La pêche dans les grands fonds, la protection de l'environnement marin et de la diversité biologique", *Annuaire du Droit de la Mer (A.D. Mer)*, Pedone, Paris, vol. 16, pp. 177–196, 2012.

[BEU 14] BEURIER J.P. (ed.) *Droits maritimes*, Dalloz, Paris, 2014.

[CAS 12] CASADO RAIGÓN R., "Le régime juridique de la protection du milieu marin dans le droit International actuel", in ANDREONE G., CALIGIURI A., CATALDI G. (eds), *Droit de la mer et émergences environnementales*, Editoriale Scientifica, Naples, pp. 21–36, 2012.

[CAZ 12] CAZALET B., FERAL F., GARCIA S., "Gouvernance, droit et administration des Aires Marines Protégées", *Annuaire du Droit de la Mer (A.D. Mer)*, Pedone, Paris, vol. 16, pp. 121–152, 2012.

[CUR 11] CURY P., MISEREY Y., *Une mer sans poissons*, Calmann-Lévy, Paris, 2011.

[DAI 02] DAILLET P., PELLET A., *Droit International public*, 7th ed., LGDJ, Paris, 2002.

[DAU 03] DAUDET Y., DUPUY P.M., COUSSIRAT-COUSTÈRE V. *et al.*, *Mélanges offerts à Laurent Lucchini et Jean-Pierre Quédeudec, La mer et son droit*, Pedone, Paris, 2003.

[DEM 09] DE MARFFY-MANTUANO A., "What International coordination for marine biodiversity governance in areas beyond national jurisdiction", in ROCHETTE J. (ed.), *Towards a New Governance of High Seas Biodiversity*, *Oceanis*, Institut océanographique Monaco, vol. 35, nos. 1–2, pp. 205–231, 2009.

[GAL 11] GALLETTI F., "Le droit de la mer, régulateur des crises pour le contrôle des espaces et des ressources: quel poids pour des Etats en développement?", *Mondes en Développement*, vol. 39-2011/2, no. 154, pp. 121–136, 2011.

[GAL 12] GALLETTI F., CAZALET B., "Matières et instruments impliqués dans la gouvernance d'une mer semi-fermée: du droit de la mer et de la situation d'indétermination des "eaux sous juridiction" en Méditerranée à l'invention des nouveaux zonages écologiques", in RÍOS RODRIGUEZ J., OANTA G.A. (eds), *Le droit public a l'épreuve de la gouvernance. El Derecho Publico ante la gobernanza*, Presses Universitaires de Perpignan, Perpignan, pp. 257–296, 2012.

[GAL 14] GALLETTI F., "La protection juridique des réseaux écologiques marins. Compétences et implications du droit de la mer contemporain", in SOBRINO HEREDIA J.M. (ed.), *La contribution de la Convention des Nations Unies sur le Droit de la Mer à la bonne gouvernance des mers et océans*, Editoriale Scientifica, Naples, vol. 2, pp. 765–791, 2014.

[GAL 15] GALLETTI F., CHABOUD C., "Aires marines protégées et résistance aux risques liés au changement climatique: une fonction rénovée pour de nouvelles politiques publiques?", in BONNIN M., LAË R., BEHNASSI M. (eds), *Les Aires marines protégées en question. Défis scientifiques et enjeux sociétaux*, IRD Editions, Marseille, pp. 77–88, 2015.

[GHE 13] GHEZALI M., "Les territoires de la Gestion intégrée des zones côtières et marines", *Vertigo, la revue électronique en sciences de l'environnement*, available at http://vertigo.revues.org/14305, special edition no. 18, 2013.

[IND 13] INDEMER, *Droit de la mer et Droit de l'Union Européenne, cohabitation, confrontation, coopération?*, Pedone, Paris, 2013.

[LEH 10] LEHARDY M., "Les AMP et la gestion intégrée des océans", *Annuaire du Droit de la Mer*, Pedone, Paris, vol. 14, p. 309–344, 2010.

[LER 14] LEROY A., Les transformations du droit des pêches face au problème de la pêche illicite, non déclarée, et non réglementée, dite pêche "INN", thesis on public Law in progress GALLETTI F. (ed.), University of Perpignan, 2014–2017.

[LÓP 82a] LÓPUSKI J. (ed.), "Prawo morskie", *Encyklopedia Podreczna prawa morskiego – Encyclopédie pratique de droit maritime*, Gdansk, Poland, 1982.

[LÓP 82b] LÓPUSKI J. (ed.), "Miedzynarodowe Prawo morza – Droit International de la mer", *Encyklopedia Podreczna prawa morskiego – Encyclopédie pratique de droit maritime*, Gdansk, Poland, 1982.

[LUC 82] LUCCHINI L., "De l'emprise à la gestion: existe-t-il une politique française de la mer?", *Journal de Droit International (JDI)*, pp. 567–623, 1982.

[LUC 84] LUCCHINI L., "La troisième Conférence des Nations Unies sur le droit de la mer face au phénomène des Méditerranées ou le triomphe de l'Etat océanique", *Droits et libertés à la fin du XXᵉ siècle, Etudes offertes à Claude Albert Colliard*, Pedone, Paris, pp. 289–310, 1984.

[LUC 90] LUCCHINI L., VOELCKEL M., *Droit de la mer, tome I, La mer et son droit – Espaces maritimes*, Pedone, Paris, 1990.

[LUC 96a] LUCCHINI L., VOELCKEL M., *Droit de la mer, tome II, volume 1, Délimitation*, Pedone, Paris, 1996.

[LUC 96b] LUCCHINI L., VOELCKEL M., *Droit de la mer, tome II, volume 2, Navigation et pêche*, Pedone, Paris, 1996.

[LUC 03] LUCCHINI L. (ed.), "Les zones maritimes en Méditerranée", *Revue de l'INDEMER*, Paris, no. 6, 2003.

[MAR 14] MARINE POLICY, "Special section: advancing governance of areas beyond national jurisdiction", *JMPO*, vol. 49, pp. 81–194, November 2014.

[MOM 95] MOMTAZ D., "L'accord relatif à la conservation et à la gestion des stocks de poissons chevauchants et de grands migrateurs", *Annuaire français de droit International (AFDI)*, CNRS, Paris, vol. 41, pp. 676–699, 1995.

[ORR 99] ORREGO VICUÑA F., *The Changing International Law of High Seas Fisheries*, Cambridge University Press, Cambridge, 1999.

[PAN 97] PANCRACIO J.P., *Droit International des espaces*, Armand Colin, Paris, 1997.

[PON 96] PONTAVICE (DU) E., RODIERE R., *Droit maritime*, Dalloz, Paris, 1996.

[PRO 07] PROUTIERE-MAULION G., BEURIER J.P., "Quelle gouvernance pour la biodiversité marine au-delà des zones de juridiction ?", *Idées pour le Débat*, no. 07/2007, Gouvernance mondiale, IDDRI, Paris, 2007.

[QUE 94] QUENEUDEC J.P., "Le nouveau droit de la mer est arrivé !", *Revue Générale de Droit International Public (RDGIP)*, pp. 865–870, 1994.

[QUE 97] QUENEUDEC J.P., "Mer territoriale et territoire maritime", *Annuaire du Droit de la Mer (A.D. Mer)*, Pedone, Paris, vol. 2, pp. 105–166, 1997.

[ROD 97] RODIÈRE R., PONTAVICE (DU) E., *Introduction au droit maritime*, 12th ed., Dalloz, Paris, 1997.

[ROS 12a] Ros N., "La Méditerranée: cas particulier et modèle avancé de gestion de la haute mer", *Annuaire du Droit de la Mer (A.D. Mer)*, vol. 16, pp. 33–62, 2012.

[ROS 12b] Ros N., "La lutte contre la pêche illicite", in Cataldi G., Andreone G., Caligiuri A. (eds), *Droit de la mer et émergences environnementales*, Editoriale Scientifica, Naples, pp. 69–122, 2012.

[ROS 14] Ros N., "Conclusions générales à la contribution de la CNUDM à la bonne gouvernance des océans", in Sobrino Heredia J.M. (ed.), *La contribution de la Convention des Nations Unies sur le Droit de la Mer à la bonne gouvernance des mers et océans*, Editoriale Scientifica, Naples, pp. 868–894, vol. 2, 2014.

[ROS 15] Ros N., Galletti F., *Quelles contributions de la Méditerranée et des mers semi-fermées au droit international de la mer?*, Editoriale Scientifica, Naples, p. 434, 2015.

[SAL 01] Salmon J. (ed.), *Dictionnaire de droit International public*, Bruylant/AUF, Brussels, 2001.

[TAS 13] Tassin V., *L'extension du plateau continental: consécration d'un nouveau rapport de l'Etat à son territoire*, Pedone, Paris, 2013.

[VIG 00] Vignes D., Casado-Raigon R., Cataldi G., *Le Droit International de la pêche maritime*, Bruylant, Brussels, 2000.

[VIN 08] Vincent P., *Droit de la mer*, Larcier, Brussels, 2008.

2

The Governance of the International Shipping Traffic by Maritime Law

2.1. Introduction

Applied to human activities at sea, governance, in the sense of "getting governance right", is meant to regulate the human and social behavior in their relationship – whether positive or negative – with the oceans. In this sense, the international shipping traffic, which involves a very large number of operators and enormous amounts of money, is undoubtedly the most symbolic activity of human ocean exploitation.

Ninety percent of worldwide commerce is travel by ship, which justifies the institution of rules, both mandatory and optional, in order to ensure the sustainability of this shipping traffic from a social, economic and environmental point of view.

Such is the subject of maritime law: "getting governance right" of the ocean industry in accordance with maritime security.

Equally, the meaning of maritime law (section 2.1.1) is determined by the necessity to take into account both the vulnerability of human communities to the dangerous nature of marine environment and the ocean vulnerability to antropogenic pressures (section 2.1.2). To assert and organize this search for balance, maritime law relies on a corpus of novel rules developed mainly at the international level (section 2.1.3).

Chapter written by Cécile DE CET BERTIN and Arnaud MONTAS.

2.1.1. *Meaning and definition of maritime law*

Because of the wide variety of legal issues having to do with marine matters, defining maritime law is not an easy task. "There are three sorts of Men", wrote Plato, "The Dead, the Living, and Those that go to Sea". For the latter, specific legal rules that have no purpose on land but are necessary at sea in order to regulate human presence there. In order to protect man from the dangers of the sea as much as the sea from the human pressures, maritime law is an original system governed by its own rules, methods and institutions. Marine law designates all legal situations (of private law) exposed to the hazards of maritime navigation, thus it is designed to answer questions whose uniqueness stems from the adversity of the marine environment.

As attested to by the recognition of ecological harm in terms of reparable damages (the "Erika" affair), maritime law has always been confirmed as precursory law. The solutions of maritime law have often been reproduced in terrestrial law. Notably, maritime transport law has given rise to legal concepts that are now reliable, such as the making paperless of transport documents, or the emergence of environmental law resulting from the black tides caused by shipwrecks. In the same vein, some noteworthy institutions of marine law that have long been ignored by common law are progressively emerging into prominence such as the limitation of responsibility, which is widespread in maritime law and now becoming known in terrestrial law. This is an example of "the sea as the mother of the law" [SCH 81]. Moreover, "in unforeseen and unsurmountable circumstances, in which general law yields and capitulates, it is the very function of maritime law to anticipate the worst" [REM 98].

2.1.2. *Fundamental principles of maritime law*

The most ancient fundamental principle of maritime law has to do with the "risk of the sea", the natural or anthropogenic risk around which the discipline was founded; "the perils of the sea pervade and shape the entire discipline of maritime law" [VIA 97]. Whether we call it danger or peril of the sea, marine hazard or fortunes of the sea, risk at sea is a reality at all times, and the evolution of maritime law has been based on the necessity of anticipating these risks and on limiting their consequences. The objective of maritime security is aimed, in this sense, at ensuring the security of the people exposing themselves to these risks, of the vessels facing them, of the environment threatened by them and, finally, the trade surrounded by them.

In general law, legal vocabulary tells us that risk is "a prejudicial event of which the occurrence is uncertain, both in terms of its happening and of the date of its happening" [COR 11]. Applied to maritime law, the concept of "risk of the sea" has motivated the emergence of a specific responsibility based in part on fault (unlike the trends in general law) and in part on the sharing and restriction of reparations in the event of damages consecutive to its occurrence. Based on the feeling of mutual dependence that has always existed among adventurers on the seas, these special rules have made maritime law into a "solidarist" discipline [VIA 97]. It is now a means of protecting both the physical selves and the heritage of those who expose themselves to risk, while also defending maritime security via better governance of conduct. This "maritime responsibility" is presented as a privilege granted to actors on an expedition; however, depending on the nature of the event, this advantage will be maintained only with regard to risks posed by the sea, and will collapse if the damages sustained are caused by human fault.

2.1.3. General sources of maritime law

The history of maritime law has been a rich one since the beginnings of its history. Its noteworthy sources include the Rolls of Oléron adopted by Eleanor of Aquitaine in around 1150, which inspired numerous texts. Later, the Consulate of the Sea, drafted in the 13th or 14th Century, dealt with the construction of seagoing vessels, the transport of merchandise and incidents at sea. In the 17th Century, the 704-article-long Great Marine Ordinance of August 1681, also called the Colbert Ordinance, had a decisive influence on the development of modern maritime law and even survived the French Revolution intact to such an extent that in 1807, volume II of the code of commerce reused the basic principles of maritime law developed during the reign of Louis XIV. Though it has now become obsolete, the Colbert Ordinance was only revoked by ordinance 2006-46 of April 22, 2006 relative to the legislative part of the general code of public sector property. Book II of the code of commerce would be modified often to take into account the evolution of maritime law and emancipation from international rules. Between 1966 and 1969, the Rodière laws[1] sanctioned the freeing of maritime law from the code of commerce. In late 2010, the coming into force of the legislative part of the transport code rationalized and standardized maritime law. Linked by a double relationship of authority and symbiosis, both international and domestic regulations serve as modern sources of maritime law.

2.1.3.1. *International sources of maritime law*

2.1.3.1.1. International conventions

International institutions play a pre-eminent role in the production of maritime law and participate actively in the internationalization of the discipline. Numerous international conventions contribute to maritime law but, while some of them constitute pillars of the legal discipline (such as the Safety of Life at Sea (SOLAS), MARPOL and STCW conventions), others are more modest in terms of their objectives or the legal system for which they supply structure. In any case, it would be unreasonable to attempt an exhaustive list of these conventions here, and we will emphasize only those concerning maritime transport activities in the following sub-chapters.

2.1.3.1.2. European Union (EU) law

First instituted as a safeguard, normative action by the EU in maritime affairs has been very widely developed and is now a driving force in a true integrated European maritime policy. European control is becoming increasingly directive in ever-widening circles, requiring Member States to apply EU regulations and to transpose these directives onto domestic law. Formerly concentrated mainly on the regulation of fishing and competition, and more precisely on the application to maritime transport of exemptions on the principle of freedom to provide services, free competition and free market access through liner conferences, the EU's actions today have grown increasingly focused on maritime security, to the extent that the Union now has its own maritime security and navigation policy as part of its transport policy. This is attested to by the adoption of numerous texts among the multiple regulatory instruments used.

---

1 Law no. 65-420 of June 18, 1966 on chartering and maritime transport contracts, JO 24 June; law no. 67-5 of January 3, 1967 relative to the status of vessels and other sea construction, JO 4 January; law no. 67-545 of July 7, 1967 relative to incidents at sea; JO 9 July; law no. 69-8 of January 3, 1969 relative to munitions and maritime sales, JO 5 January.

Following the Erika disaster in December 1999, the European Commission grouped a number of measures designed to improve maritime security into three "packages", called "Erika I, II and III". Several directives were adopted in application of these legislative packages, among them the directive of June 27, 2002 on the monitoring system for the shipping traffic, the directive of April 23, 2009 concerning the port state control and mandatory insurance for shipowners for maritime claims, and the directive of October 21, 2009 relative to pollution caused by vessels and to the introduction of penalties for infringements of international law.

2.1.3.2. *Domestic sources of maritime law*

Despite the international quality inherent in maritime law, the teachings of comparative law have always shown the pluralism of maritime legal cultures. In particular, despite having experienced its golden age in the second half of the 19th Century, the contribution of English Common Law to maritime law remains significant today. Long divided up within several codes and other scattered laws, French maritime law was the object in 2010 of a double operation to rationalize and standardize it within the transport code, which is certainly not a true maritime code. Consequently, ordinance no. 2010-1307 of October 28, 2010[2], used in application of article 92 of law no. 2009-526 of May 12, 2009, relative to the simplification and clarification of the law and the streamlining of procedures, resulted in the creation of the legislative part of the transport code, which became effective on December 1, 2010. More recently, ordinance no. 2011-635 of June 9, 2011[3] rendered French law compliant with the maritime security objectives of the EU, by adapting the legislative part of the code to the directives stated in the package Erika III. These provisions are particularly concerned with the strengthening of port state controls, the standardization of investigation procedures after accidents, increased monitoring of marine classification companies and the prevention of maritime disasters.

---

2 JO 3 November. Ordinance no. 2011-204 of February 24, 2011 including various provisions for the adaptation of the transport code to European Union law and to international conventions in the fields of transport and maritime security (JO 10 June) has contributed various modifications to the ordinance of 28 October 2010 in order to reaffirm established law and to clarify certain provisions subject to overly broad interpretations (JO 25 February).
3 JO 10 June 2011.

The transport code includes more than 2,200 articles and is composed of six parts[4], the fifth and longest part is devoted to "Maritime transport and navigation[5]". This part is made up of seven books: Book 1: Vessels; Book II: Maritime navigation; Book III: Maritime ports; Book IV: Maritime transport; Book V: seafarers; Book VI: French international registry; Book VII: Provisions relative to overseas territories.

Though part five incorporates a number of scattered maritime laws, notably the Lois Rodière of 1966, 1967 and 1969, it does not cover all maritime issues, some of which are still addressed by other codes. For example, maritime mortgages and the nationality of vessels are regulated by the customs code except in specific cases. Marital status and wills on board vessels remain under the civil code. Maritime insurance still falls principally within the remit of the insurance code, and to a lesser extent of the environmental code. The suppression of acts of maritime piracy, recently updated, is contained in the criminal code, the code of criminal procedure, and the defence code. Some questions relative to the local organization of maritime transport and to nautical leisure activities are addressed by the general code of territorial governments and the sports code. Since 2007, laws relative to both professional and recreational sea fishing have been part of the rural code. Provisions having to do with the marine environment, protection of the coastline and responsibility for pollution by hydrocarbons are contained mainly in the environmental code. Finally, submarine archaeological wrecks are part of the heritage code. As it now stands, the transport code, though incomplete, constitutes a significant advance in maritime law through the standardization and defragmentation of maritime laws it enacts.

In substance, part five, which is principally standardized to established law, does contain some new facets of varying scope and range. In particular, article L.5000-2 of the transport code contributes to French law, which did not previously include it, a legal definition of a vessel: "1. Any floating craft built and equipped for maritime navigation for the purposes of trade, fishing or recreational activities, and appointed for these purposes; 2. Floating

---

4 www.legifrance.gouv.fr/affichCode.do?cidTexte=LEGITEXT000023086525&dateTexte=20111111.

5 Part one contains provisions having to do with all forms of transport; part two addresses rail transport; part three addresses road transport; part four addresses domestic navigation and river transport. Part five addresses maritime transport and navigation, and part six deals with civil aviation.

vehicles built and equipped for maritime navigation, appointed for public service of an administrative or industrial and commercial nature". This generic definition, which does not answer questions concerning the qualification of certain floating crafts, is a cross-sectoral definition that contrasts with the circumstantial definitions belonging to international conventions which define vessels precisely according to their object and the conditions of navigation proposed for each.

Upstream, the governance of maritime shipping traffic, a major part of maritime law, relies on a number of legal instruments contained in the institutions and sources that determine it (section 2.2).

Downstream, using these instruments for the regulation of conducts, maritime practices have put in motion a large number of contracts participating, at their own levels, in the governance of maritime transport by law (section 2.3).

2.2. Legal instruments of governance: institutions and sources of maritime transport law

The development of maritime transport law is part of the governance of the seas, as maritime transport is a vital economic activity in the international merchandise trade. The necessity of connecting continents and the power of merchant vessels has made maritime transport pre-eminent in this trade. Compared to planes, which can fulfill the same purpose, it has been observed that for the transport of half a million tons of oil to Europe from the Persian Gulf, a seagoing vessel would require 2 months while it would take the largest airplane available 2 years. Maritime transport, then, is relatively rapid [VIG 87]. The major event in this activity in the past 20 years has been the rapid growth in the transport of various types of merchandise by container ships. In 2012, traffic in container shipping worldwide increased by 5.9%, reaching 572.8 million 20 foot equivalent units (TEU)[6], and maritime traffic overall reached 8.7 billion tons[7].

---

6 The 20 foot equivalent (TEU) is the standardized measurement unit for containers. A standard container of 1 TEU measures 2.591 m (8.5 feet) high by 2.438 m wide (8 feet) and 6.096 m (20 feet) long, representing around 38.5 cubic meters.

7 UNCTAD, 2012 study on maritime transport, p. 16.

The preponderance of maritime transport in the international merchandise trade explains the fact that maritime law is instituted in large part on an international scale, and that economic organizations for regional integration, such as the EU, have made it one of the areas in which they are competent to act. The main legal instruments of governance are these international institutions created to develop rules common to the states that wish it (see section 2.2.1). However, from the European point of view, belonging to the Union, which adopts its own regulations, also generates a framework for this activity (see section 2.2.2).

2.2.1. *Development of international regulations*

There are several international organizations within which the maritime transport regulations that make up international law are conceived and debated and then adopted, where applicable.

2.2.1.1. *Origins of international rules*

International organizations have their origins in a multilateral treaty that may be referred to as a convention, pact, set of statutes or constitution. This constitutive act establishes the legal character of the organization and its capacity to act in a certain domain with certain means and according to a certain mode of operation.

Organizations whose remit involves acting in matters of maritime transport include the International Maritime Organization[8] (IMO) and the International Labor Organization[9] (ILO). There are also organizations which, because they contribute more broadly to the governance of global trade, may influence maritime commerce. This is true for the World Trade Organization (WTO)[10]. Other institutions, such as the United Nations Conference on Trade and Development[11] (UNCTAD) and the United Nations Commission on International Trade Law[12] (UNCTIL), develop material regulations that govern

---

8 International Maritime Organization (IMO): www.imo.org.
9 International Labour Organization (ILO): www.ilo.org.
10 World Trade Organization (WTO): www.wto.org.
11 United Nations Conference on Trade And Development (UNCTAD): www.unctad.org.
12 United Nations Commission on International Trade Law (UNICITRAL); voir www.unicitral.org.

the relationships between maritime transport operators (mainly shipowners, carriers and shippers[13]).

2.2.1.1.1. International Maritime Organization

The convention creating this United Nations organization was adopted in Geneva on March 6, 1948. At the time, it was called the Intergovernmental Maritime Consultative Organization (IMCO), but subsequently changed its name to become the International Maritime Organization (IMO).

The IMO includes 170 member states, three of which have been associated members since June 2013. It is headquartered in London, England, with its governing body, the Assembly, meeting every 2 years. Between sessions of the Assembly, a Council composed of 40 governments elected by the Assembly acts as the governing body. In December 2013, during the 28th session of the Assembly, the 40 member states of the Council were elected for the 2014–2015 period, divided into three categories: A, B and C.

Category A is made up of the 10 countries with "the greatest interest in supplying international maritime navigation services". These are China, Greece, Italy, Japan, Norway, Panama, the Republic of Korea, Russia, the United Kingdom and the United States. Category B is composed of countries "with the greatest interest in international maritime trade". Finally, category C includes "countries with a particular interest in maritime transport or navigation and whose election to the Council will ensure that all the world's major geographical regions will be represented". In the most recent election, Egypt left the Council, which consequently no longer includes a representative for the countries of the Middle East.

Aside from these two principal bodies, the IMO carries out its work through several committees and subcommittees. These include the Maritime Safety Committee (MSC), which deals with all issues relative to the security of maritime transport, and the Marine Environment Protection Committee (MEPC), which coordinates actions in the field of prevention and control of environmental pollution caused by ships.

---

13 A shipper, in maritime law, contracts with a maritime carrier for the delivery of merchandise from one port to another. It is important to understand that this is not the party that carries out the loading of a vessel, a physical operation carried out by a cargo handling company. See *infra*, section 2.3.2.

2.2.1.1.2. International Labor Organization

The origins of the act creating the ILO lie in the 1919 Treaty of Versailles, of which it formed Part XIII. It was subsequently separated to become the Constitution of this international organization, which now includes 185 member states. In 1944, a declaration of the fundamental goals and objectives of the ILO was adopted. This document, called the Philadelphia Declaration, sets out the founding principles on which the policies of member states are based, and was subsequently incorporated into the Constitution. The fundamental principles of the ILO are as follows: (1) labor is not merchandise; (2) freedom of expression and association is a vital condition for sustained progress; (3) poverty, wherever it exists, poses a danger to the prosperity of all; and (4) the fight against want must be conducted with unremitting energy within each nation and via ongoing and concerted international effort, in which representatives of workers and employers cooperate on an equal footing with those of governments and participate in free discussion and decision-making of a democratic nature with a view to promoting the common good. The ILO is headquartered in Geneva, Switzerland.

The permanent organization includes three governing bodies; a general conference of representatives of member states; a board of directors and the International Labor Office, which is under the leadership of the board of directors (article 2 of the Constitution). This board of directors is composed of 56 individuals, 28 of whom are representatives of member governments, 14 of whom represent employers and 14 of whom represent workers. For this reason, the ILO is said to include tripartite representation.

2.2.1.1.3. World Trade Organization

The World Trade Organization differs from the organizations discussed before; it is intended as a space for multilateral negotiations on questions of trade. It was conceived as part of the GATT[14] during the international negotiations of the Uruguay Round, begun in Punta del Este in September 1986 and completed in Marrakesh in 1994 (the Uruguay Cycle).

The World Trade Organization does not possess decision-making bodies such as a Council and a Board of Directors, like the two organizations above do. It is overseen by the governments that are its members, of which there

---

14 General agreement on tariffs and trade.

were 159 in 2014. Decisions are made by all members, either by Ministers within the ministerial conference, which meet every 2 years, or by the ambassadors and delegates who meet regularly in Geneva. Decisions are made by consensus within the organization, meaning there are no voting procedures. A majority vote is possible by agreement among members, but this does not occur in practice.

The accord establishing the WTO provides for a general Council that carries out three functions; that of a board of supervisors acting on behalf of the ministerial conference for all matters falling within the scope of competence of the WTO; that of a body for the settling of disputes which oversees the implementation of procedures to settle trade disputes between countries; and that of an examining body for the commercial policies of member states.

This general Council, which is made up in principle of the ambassadors of its member states and the heads of their delegations, includes three subcouncils, which oversee trade in services, trade in merchandise and the aspects of intellectual property law that touches on trade (Trade-related Aspects of intellectual Property Rights (TRIPs)). Only the first of these deals with maritime transport, which is a service activity. This specialized sub-council is particularly concerned with the General Agreement on Trade in Services (GATSs). In addition to these bodies, various committees, work groups and experts contribute to the work of the WTO. Finally, a secretariat located in Geneva supplies technical support to the councils and other committees and to the ministerial conferences. It also supplies legal assistance in the settling of disputes and advises governments wishing to become members of the WTO.

2.2.1.1.4. United Nations Conference for Trade and Development

The United Nations Conference for Trade and Development (UNCTAD) is a subsidiary body of the United Nations created to handle claims made by developing countries in the early 1960s. The first conference was held in Geneva in 1964, and became an institution shortly thereafter. It is headquartered in Geneva and includes 194 member states. It is responsible for dealing with questions relative to trade and economic development within the United Nations system.

The "conference" is its governing body, held every 4 years. The Trade and Development Council meets between two conferences. The institution also includes three commissions: Trade and Development; Investment, Enterprise and Development; and Science and Technologies for Development. The secretariat provides operational and technical services to intergovernmental bodies and is overseen by a secretary-general.

UNCTAD has produced a summary and analysis of maritime transport activity worldwide every year since 1968. These economic and legal observations are contained in a publication entitled "Review of Maritime Transport", which contains a wealth of information on the evolution of international maritime traffic, its structure, the system of ownership and registration of the global fleet, the state of supply and demand in maritime transport worldwide, the shipping market, the status of ports, and legal questions; that is the evolution of legislation in maritime transport activity.

2.2.1.1.5. United Nations Commission on International Trade Law

The United Nations Commission on International Trade Law (UNCITRAL) was created in 1966 via a UN General Assembly resolution authorizing it to foster the harmonization and modernization of international trade law.

The members of UNCITRAL are chosen from among UN member states representing various legal traditions and levels of economic development. In 2002, the number of Commission members was raised to 60: 14 African nations, 14 Asian nations, 8 nations from Eastern Europe, 10 from Latin America and the Caribbean, and 14 nations from Western Europe and other regions. Members are elected for a term of 6 years, and the mandate for half of them expires every 3 years.

The work of UNCITRAL is carried out at three levels. The first level is that of the Commission itself, which holds a yearly plenary session alternating between Vienna and New York. The second level includes intergovernmental work groups which are responsible for developing the topics included in the Commission's work programme. The third level is that of the secretariat, located in Vienna, which provides operational assistance to the Commission and work groups.

The Commission is endowed with a "bureau" at the start of each of its sessions; this bureau is elected from among the 60 members for a term lasting until the start of the next annual session. It is composed of a president, three vice-presidents and a rapporteur, representing each of the five regions from which the members of the Commission originate.

During its first session in 1968, the Commission chose nine subjects as the basis of its work program; these include transport and, as we will see below, instruments able to be used for the benefit of international governance of maritime transport were created in this context.

Along with UNCTAD, the WTO, the ILO and the IMO contribute to the development of international regulations; these do not all have the same scope and they are not all of the same type, but all play a role in the formation of a system of international maritime transport law.

2.2.1.2. *International maritime transport law*

International maritime transport law participates, obviously, in the governance of the seas. Taken as a set of regulations that are international in origin and are not contained within the legislation of a single country, but rather in a text adopted by several of them, it is manifested through the work of various institutions that have developed regulations applicable to it. These rules form a framework for maritime transport activity, with their object being maritime transport markets. They may also be material regulations applicable to ships and their navigation, the work of marine operators, or contracts concluded for the conveyance of individuals or merchandise. As the second part of this study is entirely dedicated to maritime contracts, we will not address the regulations concerning these contracts here; this leaves us with the regulation of the markets in which ships carrying out transport activities operate, and regulations applicable to ships and their navigation, as well as labor that takes place on board.

2.2.1.2.1. Regulation of maritime transport markets

There are multiple maritime transport markets, due to the specialization of ships for the transport of various specific types of merchandise, there exist multiple maritime transport markets. These specialized ships (tankers, grain carriers, gas transport vessels and other freighters) carry out transport on demand (tramping). They are chartered by traders in raw materials. Freight costs, that is the costs of chartering[15] ships agreed upon in order to carry out

---

15 See *infra*, section 2.3.1.

the transport of merchandise, fluctuate mainly due to economic conditions and are subject to the law of supply and demand. In 1998, a memo from the Trade Services Council of the WTO made this revealing observation on the functioning of these markets: "for the most part, the transport of freight (crude and refined oil, iron ore, grain, coal, and bauxite), which represents 67.7% of the total traffic volume, is not subject to any restriction except in the case of one or two countries. It is organized like a cash market (there is also a forward market), and markets are allocated in a highly competitive manner, on the basis of the lowest freight costs" (S/C/W/62, November 16, 1998, p. 2).

Regulated markets are those of regular shipping services, which currently include a large number of container ships. Regulations have been established, notably in order to avoid negative effects on free competition, as in these markets, shipowners and other ship operators sometimes form groups to offer their transport services[16].

Aside from these market regulations, international maritime transport is affected by multilateral negotiations held under the aegis of the WTO. The cycle of negotiations having to do with services trade began in 2000. Subsequent to a ministerial declaration adopted at the Hong Kong Conference in 2005, requests have been made by some members in the domain of maritime transport. These requests concern the elimination of reserved portions of cargo and restrictions relative to foreign participation in shareholding and the right to establish a commercial presence for the international transport of merchandise, and for services secondary to maritime transport, such as those related to the handling of merchandise.

In this context, a general accord on services trade based on three pillars was adopted. The first of these pillars is a framework agreement containing fundamental obligations applicable to all member states. The second concerns lists of commitments established by countries, which proclaim other specific national commitments requiring an ongoing process of liberalization. The third pillar is composed of a number of appendices which address situations proper to this or that sector of services. Maritime transport is not the subject of an appendix, which shows a certain difficulty for member states in accepting the liberalization of maritime transport; however, the sector is concerned by the appendix relative to financial services

---

16 For European regulations, see *infra*, section 2.2.2.

(banking and insurance services). A memorandum of agreement on commitments relative to financial services specifies that each member state will allow non-resident suppliers of financial services to provide, under the conditions granted to residents (national treatment), insurance services against risks related to maritime transport. In other words, and in the manner of the free provision of services, it is specified that the host state will allow foreign suppliers of insurance services for maritime transport to benefit from the same conditions of execution of their activity as its nationals.

2.2.1.2.2. Regulation of ships and maritime navigation

The International Maritime Organization establishes international regulations principally in matters of maritime security and the protection of the marine environment. These include a remarkable number of regulations pertaining to ships and maritime navigation. To cite just the principal conventions, we would note three major international conventions: SOLAS, MARPOL and STCW.

The first of these conventions, SOLAS, was established in 1974 in its current version, and has been extensively augmented and amended since then. It followed two previous conventions; the first, in 1914, was established by an international community grieving the 1912 sinking of the Titanic. Adopted shortly before the 1st World War, it remained in abeyance and a new SOLAS convention was adopted in 1948. This was followed by a third convention in 1960. Today, SOLAS stands as a monument in maritime law. It includes regulations pertaining to the safety of ships and navigation, as well as international regulations intended to prevent collisions at sea (COLREG, for Collision Regulation), as well as the International Safety Management (ISM) code and the International Ship and Port Facility Security (ISPS) code. The latter was debated and subsequently adopted following the attacks on New York of September 11, 2001. Protection against external threats to ships (safety), had become a major concern for the United States, and the ISPS code was adopted in response to these concerns.

The International Convention for the Prevention of Pollution from Ships, called MARPOL, is fully as important as the preceding convention, and has been the subject of numerous developments, the principal objective of which is to protect the sea and marine environment from harm that may be caused by ships (see *infra*, Chapter 3, on marine pollution).

The last IMO convention that can be cited for its importance in matters of maritime security – though these three conventions do not constitute the whole of the IMO's regulatory work – concerns training standards for seafarers; the issuance of certifications; and watchkeeping of ship's crews; it is known as the STCW[17]. It includes a double set of regulations; one pertaining to the minimal requirement states must fulfill in order to be granted professional certifications for sailors, and deck watch, the other pertaining to watchkeeping on board ships (engineering watch etc.). This convention has to do partly with work on board ship (watcheeping) which is also the subject of regulations adopted by the ILO.

2.2.1.2.3. Regulation of labor on board ships

Since its creation, the ILO has adopted 396 legal instruments that can be grouped as follows: 189 conventions, five protocols and 202 recommendations[18]. Some of these instruments have to do with maritime labor, including the 2006 Maritime Labor Convention (MLC)[19], which incorporates the standards contained in previous MLCs as well as the fundamental principles set forth in other international labor conventions (see the convention's Preamble). Article X of the MLC concerns the 37 previous conventions adopted between 1920 and 1996.

The MLC, which went into effect in August of 2013, is part of a codification of international maritime law. Its structure is complex to such an extent that it includes an explicative note which is not part of the convention itself, but is intended to facilitate its reading. This structure is composed of articles and regulations that set forth its fundamental rights and principles as well as the basic obligations of states ratifying the convention, which can be modified only by a conference of member states (Article XIV of the convention). It also includes a code, which indicates how regulations must be applied. This code itself is composed of two parts: Part A, which includes the required standards, and Part B, which lists optional guiding principles. This code can be modified more simply than the rules and articles (article XV of the Convention), but these modifications cannot alter the general impact of the articles and rules.

---

17 The acronym stands for Convention on Standards of Training, Certification and Watchkeeping for Seafarers.
18 Information supplied by the ILO Website
19 Acronym for *Maritime Labor Convention.*

The provisions of the regulations and the code are grouped into five categories, as follows:

– category 1: minimal conditions required for labor by seafarers aboard ships;

– category 2: conditions of employment;

– category 3: accommodation, free time, meals and table service;

– category 4: health protection, medical care, well-being and protection in matters of social security;

– category 5: compliance and application of provisions.

Each of these categories contains the rule, the mandatory standard (A) and the guiding principle (B). Thus, the first regulation in category 1, rule 1.1, is read with mandatory standard A1.1 and guiding principle B1.1 (minimum age for work on board a ship).

This complex construction is the result of the authors' desire to encourage compromise around the text so that it would be ratified by as many states as possible. The various prior conventions had seemed too rigid, and did not win a great deal of confidence on the part of nations, which did not rush to ratify them. The flexibility introduced by the new text, particularly with its non-mandatory guiding principles, as well as the ability to revise required standards more rapidly, compensates for the apparent complexity of the convention's structure.

As with all international conventions, its effectiveness is dependent on the states that must ratify it and comply with its terms. This convention became effective on August 20, 2013, after at least 30 member states of the ILO, representing a total of 33% of the world fleet had ratified it (this was a condition of its becoming effective). Of the 30 states that necessarily ratified the treaty, 15 are members of the EU[20], a proportion that shows the influence the EU can have on governance in the field of maritime transport. Furthermore, shortly after the convention became effective, two directives of

---

20 See the Commission's report to the European Parliament and to the Council on the application of 2009/21/CE concerning compliance with obligations by flag states, COM(2013) 916 final, p. 13.

the European Council and Parliament were adopted[21] in favor of the incorporation of this new convention into the EU's rules of law.

2.2.2. European maritime transport regulations[22]

The EU is a singular international organization[23]. It belongs to the category of regional economic integration organizations, of which it is a unique example, having been particularly and highly perfected in comparison to its counterparts[24]. However, according to the so-called Principle of conferral, and like any international organization, it has no competence other than what is attributed to it by its member states. These competences are evolving not only in material terms, but also with regard to their implementation, under the effects of successive modifications of the EU's constitutive treaties. These determine the EU's field of action, and the details of how it may act. This justifies a prior examination of the Union's competences in matters of maritime transport before we present its actions in the matter.

2.2.2.1. Statement of UE[25] competence in the area of maritime transport

If we consider the subject from the law determined by the Treaty of Lisbon, the extent of the Union's competences must be measured only in comparison to those of its member states. The Treaty on the Functioning of the European Union (TFEU) sets general rules on this point in article 3 to 5

---

21 Directive 2013/38/UE of the European Parliament and the Council of August 12, 2013 modifying directive 2009/16/CE pertaining to state control of ports, JO L 218 of August 14, 2013, pp. 1–7. Directive 2013/54/UE of the European Parliament and Council of November 20, 2013 relative to certain responsibilities of flag states in matters of compliance with and application of the Maritime Labor Convention, 2006, JO L 329 of December 10, 2013, pp. 1–4.

22   All   European   Union   regulations   can   be   accessed   at   www.eur-lex. europa.eu.

23 For a general overview, see www.europa.eu.

24 There are free trade associations or economic communities on every continent, but these do not have the same degree of integration as the European Union.

25 The European Union cannot be considered as the author of maritime transport regulation until the entry into force of the Treaty of Lisbon in 2009. Previously, strictly speaking, it was the European Community (EC), and before that the European Economic Community (EEC).

(the distribution of competences between the EU and its member states). Thus, there are:

– exclusive competences, meaning that the EU is the only body with the power to codify and adopt binding acts in these areas. The role of the member states is thus limited to the application of these acts, unless the Union authorizes them to adopt certain acts themselves (art. 3, TFEU);

– shared competences, when the EU and its member states are authorized to adopt binding acts in other areas. However, in these cases, the member states cannot exercise their competence except insofar as the EU has not or has decided not to exercise its own (art. 4, TFEU);

– supporting competences, when the EU can intervene only in order to support, coordinate or complete the action of its member states. It, therefore, has no legislative power in these areas, and cannot interfere in the exercise of these competences, which is reserved for member states (art. 6, TFEU).

In matters of maritime transport, shared competences were initially timidly pronounced. Indeed, the derogatory system to which maritime and aerial navigation were subject in the original treaty (the 1957 Treaty of Rome) – because the Council, composed of heads of state and governments, was the sole legislator and had a monopoly of action in the initial version of the treaty – granted the European Economic Community (EEC) limited competence in matters of maritime transport. This limitation was a condition of adoption of the treaty establishing the EEC in 1957. During their negotiations, in fact, the states were highly reluctant to agree to the transfer of their competence in matters of maritime and aerial transport; maritime authority and air authority were, therefore, considered – and still are, but to a different extent – prerogatives of state power. It should be noted that these activities, which are carried out mainly in international spaces, are conducted outside the territories of member states.

Under the terms of the treaty establishing the EEC, maritime transport was first considered as a specific service activity. Likewise the current treaty, the TFEU, is aimed at the free provision of services but differentiates the case of transport. Maritime and aerial navigation are considered separately from other modes of transport (art. 84, then 81, then 85, then 100, §2). This gives rise to a double special treatment of maritime transport: first, the special treatment of its legal situation as a service activity, and then the special treatment of it as a mode of transport.

The power of the Council (that is of heads of state and governments) to adopt measures proper to maritime navigation was not exercised during the creation of the EEC; it was not until 1986 that significant texts integrating maritime transport in the European Community were adopted, and the EU's actions in matters of maritime transport became part of this process of evolution.

2.2.2.2. *European Union actions in matters of maritime transport*

The reasons for which maritime transport occupies the place we have just described in the establishing document of the EU (the EEC treaty) lie with the European Court of Justice, which established rules in interpretation of the treaty. The EEC began its activities by adapting the rules of the Treaty of Rome to the maritime transport sector. This was followed by actions that can be considered as making up a maritime transport policy.

2.2.2.2.1. Submission of maritime transport to the general rules of the EEC treaty

Though the treaty had referred to appropriate acts that should be adopted by the Council, in matters of transport and for maritime (and aerial) navigation, in 1974 the Court of Justice, interpreting the original provisions, specified that: "Maritime transport belongs in the same category as other modes of transport, subject to the general regulations of the EEC treaty[26]". In this case, it was a matter, known as the French Seamen's Case, of determining whether the free movement of laborers applied to seamen. France reserved employment on board ships flying the French flag to French nationals at that time, but given the terms of the EEC treaty, which prohibited discrimination based on nationality and set forth a rule decreeing the free movement of individuals, was this type of restriction of employment to French nationals (and not EEC nationals) in compliance with the treaty? The Court of Justice ruled that it was not, and specified that maritime transport was not excluded from the field of application of the general rules of the EEC treaty. By this, it meant that maritime transport, despite its singular status in the treaty, was subject to the rule of non-discrimination

---

26 CJCE 4 April 1974, Aff. 167/73, Commission versus Republic of France, Rec., p. 359.

based on nationality, to the free movement of workers, and to open competition.

2.2.2.2.2. Adoption of rules appropriate to maritime navigation

The rules appropriate to maritime navigation required by article 84 §2 of the EEC treaty (currently art. 100 §2 of the TFEU) were adopted in 1986 in what was called the Brussels package. The principal contribution of this package of four regulations was the application to the maritime transport sector of the treaty's rules of competition and the principle of the free provision of maritime transport services among member states and between member states and third-party countries (international traffic).

The application of rules of competition was then subject to a specific system with Council regulation (EEC) no. 4056/86, which was repealed in 2006. The sector is now subject to general rules of competition and is not affected by specific rules except in matters concerning *consortia*, or agreements between shipowners on regular lines. These cooperation agreements for the operation of maritime transport lines are defined by the European Commission as an "agreement or a series of separate but connected agreements between line maritime companies having to do with the operation of a joint service by the parties. The legal form of these agreements is less important than the underlying economic reality; that is, the provision of a joint service by the parties[27]". This provision is aimed at preventing anti-competitive agreements in the sector.

The rule relative to the application of the principle of the free provision of international maritime transport service[28] is still in effect. It orders the application of the free provision of this service and prohibits certain restrictions of this freedom that existed at the time of its adoption. These restrictions can notably be found in the legislation of member states reserving maritime shipments or traffic to vessels flying their flag. The rule was initially applicable only to international transport, and it was not until 1992 that the rule was adopted which decreed the free provision of services

27 Regulation (EC) no. 906/2009 of the Commission of September 28, 2009 concerning the application of article 81, paragraph 3 of the treaty to certain categories of accords, decisions and practices conducted in concert between line maritime companies, JO no. L 256 of September 29, 2009, p. 31. The Commission was authorized to adopt this rule for the implementation of regulation (CE) no. 246/2009 of the Council of February 26, 2009.

28 Regulation (EEC) no. 4055/86 of the Council of December 22, 1986, JO no. L 378 of December 31, 1986, p. 1.

within a member state[29]. This text, like the preceding one, ordained that member states must not restrict access to internal maritime traffic and coastal maritime navigation to the detriment of nationals of other member states or vessels flying the flags of these states.

2.2.2.2.3. Developments of European Community actions and subsequently the European Union as part of a maritime transport policy

Developments in actions by the EU in matters of maritime transport can be distinguished according to whether these matters are considered as transport activities, in which case they belong to transport policy, or whether they are considered to be maritime activities, in which case they are part of integrated maritime policy. This is an approach introduced by the European Commission in 2007[30]. However, initially and fundamentally, maritime transport is an element of transport policy. European governance of this sector is discernable in the Commission's communication defining the strategic objectives and recommendations concerning the EU's maritime transport policy through 2018[31].

This act, issued by the European Commission, presents maritime transport as having been one of the principal elements in European economic growth and prosperity throughout its history. In it, maritime transport services are considered vital to the economy and to businesses participating in global competition. In this communication, the Commission lists the areas in which resources may be deployed; these include markets, human resources, environmental protection, safety, security, surveillance and watchkeeping, short sea shipping, and technological innovation. This vast program is also ambitious because it purports to "promote safe, secure, and efficient intra-European and international maritime transport on the seas and oceans, the long-term competitiveness of maritime transport and its related sectors in global markets, and the adaptation of the maritime transport system as a whole to the challenges of the 21st Century[32]".

---

29 Regulation (EC) no. 3577/92 of December 7, 1992 concerning the application of the principle of free movement of maritime transport services within member states (maritime coastal navigation), JO no. L 364, p. 7.

30 Blue book on integrated maritime policy, COM (2007) 575 final, October 10, 2007.

31 COM (2009) 008 final, January 21, 2009.

32 COM(2009) 008 final, January 21, 2009, §8.

2.3. Legal results of governance: maritime contracts

A proper governance of the seas means a good management of maritime relationships among individuals. From this perspective, as a formidable pathway for trade of all kinds, the sea is a privileged vector for human activities, given that 90% of global trade is conducted via ship. It is clear, then, that the economic importance of the oceans justifies the implementation of legal regulations designed to govern human activities that take place at sea, and to provide legal support for maritime contracts.

Very broadly speaking, maritime transport activity involves two principal contractual forms. The first form is the maritime chartering contract, which involves a ship made available to a shipper by a shipowner for use at sea; the other form is the transport contract, involving merchandise entrusted to a transporter by a loader for conveyance by ship. If a maritime chartering contract is mainly the result of contractual decisions made by the parties to it, a contract for the transport of merchandise or people by ship is governed by mandatory provisions, the content of which is not determined by the decisions of the parties to the contract. Finally, as the legal figure ensuring the assumption of the consequences of risks posed by sea travel, a specific place must be reserved for maritime insurance, which is considered a privileged vector of the governance of oceans.

2.3.1. *Maritime chartering contracts*

Maritime chartering is the contract by which a lessor agrees to make a ship available to a charterer in return for payment (C. transp., art. L.5423-1). The owner of the ship does not use this ship as a means of transport, therefore, but rather puts it at the disposal of a charterer in return for monetary compensation.

There are three main types of chartering: voyage chartering, in which a shipowner makes a ship available to a charterer for the transport of a given type of merchandise from one port to another; time chartering, in which a shipowner makes a ship available to a charterer for a predetermined period of several months or years; and finally bareboat chartering, in which the shipowner makes the ship available to a charterer for a predetermined period, but in this case the ship is not fitted out and lacks equipment, or is incompletely fitted out or has incomplete equipment.

In these matters, legal provisions are secondary to the will of the contracting parties (C. transp., art. L.5423-1, pgh. 2); therefore, it is only when these parties' wishes have not been expressed, or have been imprecisely expressed, that the law is applied. This is why the majority of maritime chartering law is contained within contractual relationships between parties.

A formalized chartering contract is properly called a charter-party. Though this charter-party must include a certain number of precise indications (components of individualization of the ship, names of the lessor and charterer, type of cargo, sites and timelines for loading and unloading, and freight cost), its content is mostly free and depends on the type of chartering contracted. In practice, charter-parties use boilerplate printed contracts supplied by shipping companies, which the parties to the contract are then free to modify and amend.

In the same sense, in compliance with article 3 of the Convention of Rome of June 19, 1980 on law applicable to contractual obligations, the law applicable to a chartering contract is that chosen by the parties. Likewise, the community regulation "Rome I" confirms the primacy of the law of autonomy. If the parties do not specify a choice, the contract is governed by the law of the country in which the shipper's usual residence is located, unless there are closer ties to another country, in which case the law of the latter is applied.

The actions specified by the chartering contract are effective for 1 year. The end point of this effectiveness varies according to the type of chartering; it may occur at the time the unloading of merchandise is completed, or at the time of the event ending the voyage in the case of voyage chartering; or it may occur when the contracted duration of the charter has expired, or at the time of the definitive stopping-point of its execution for time and bareboat chartering.

In determining the respective prerogatives and obligations of the shipowner and the charterer, a distinction is generally made between the nautical and commercial management of the ship leased.

While nautical management has to do mainly with the direction of the ship and its maritime fitness (costs of fitting out and maintenance of the ship, crew wages and hull insurance contract), commercial management has to do more specifically with the cargo transported as well as costs related to travel (hold management, piloting costs, and payment of taxes and entitlement to stopover in ports of call). From this point of view, the three categories of chartering evoked by the law are distinguished by whether these powers of management and nautical and commercial responsibilities lie with the shipowner or the charterer; in matters of voyage chartering the shipowner exercises and assumes nautical and commercial management simultaneously, while in time chartering the shipowner is responsible for nautical management and the charterer is responsible for commercial management, and in bareboat chartering the charterer is wholly responsible for both nautical and commercial management.

2.3.2. *Maritime transport contracts*

While chartering contracts have to do with a ship that will be used to move merchandise, maritime transport contracts are concerned with merchandise that will be moved by ship. Under the terms of these contracts, a loader agrees to pay freight costs and a transporter agrees to transport a given amount and type of merchandise from one port to another. In reality, this type of contract involves three parties: the loader who is sending the merchandise, the transporter who moves it and finally the recipient, who – even though a third party in the contract – will take delivery of it and thus benefit from legal action taken against the transporter in the event of damage to or loss of the merchandise.

After having been part of the *Ordonnance de la Marine* of August 1681 and then the code of commerce of 1807, governance of maritime contracts now falls within a remit strictly delineated by several obligatory and directly applicable international conventions. Though these conventions set forth rules that are substituted for domestic rules when transport is international, the weakness of the system arises from the heterogeneity of these sources, which constitute a mosaic of texts instituting a large number of regimes that differ subtly from another, and with no clear connections.

However, the international community has attempted to unify maritime contract law through several international conventions:

– Widely ratified, the Brussels Convention of August 25, 1924 provided for the unification of certain regulations having to do with freight bills (called La Haye-Visby rules), with the particular intention of settling conflicts between the laws of contracting states. Excluding the transport of living animals and regular carriage on deck (which includes all merchandise loaded onto the deck of a ship), it brought about compromise between the respective interests of loaders and shipowners. Considered as common international maritime transport law, it places a presumption of public responsibility on the transporter. In counterpart to this system, which is aimed at avoiding probationary difficulties, transporters may free themselves from responsibility by proving one of the 17 reasons for exoneration enumerated by the text, and benefit particularly from a legal limitation of responsibility in terms of reparation for damages for which they are accountable.

– The Hamburg Convention of March 30, 1978, having to do with the transport of merchandise by ship (called the Hamburg rules), was developed under the aegis of UNCTAD, under the influence of developing countries (loading countries). This text maintains and reinforces the system of responsibility of the transporter, which is not entitled here to cases of exemption. The Hamburg rules specify that the transporter is presumed to be at fault for damages resulting from losses or damages sustained while the merchandise was under its care, unless it can prove that every measure that could reasonably have been required for the avoidance of losses or damages was taken. It is estimated that around 5% of maritime trade worldwide is subject to the Hamburg rules.

– The Rotterdam convention on contracts for the international transport of merchandise partly or entirely by ship (called the Rotterdam rules) was adopted by UNCTAD and then by the General Assembly of the United Nations on September 23, 2009 but has not yet become effective. Its main contribution lies in the field of application of the new rules. As a multimodal convention, it is applicable to all transport preceding or succeeding maritime transport, whatever its type (road, rail or air transport). Governing international transport at the starting point or destination of a contracting country, its provisions are intended to make the legal system of merchandise transport including an international maritime phase uniform, as well as to modernize maritime transport by taking into account recent developments in the sector (electronic transport documents and containerization). At its core, the convention is quite heavily dominated by contractual freedom, which calls into question the historically imperative tendencies of regulations applicable to maritime transport contracts. Loaders must hand over merchandise in an appropriate manner, provide transporters with the information, instructions and documentation necessary for its delivery, supply transporters with the information necessary for the drafting of contractual data; and, where applicable, make the required declarations pertaining to dangerous merchandise. The loader is responsible with regard to the transporter if the latter can prove that any losses or damages sustained are the result of a failure by the loader to fulfill its obligations. Unless otherwise provided the transporter is required to transport the merchandise to its destination and deliver it to the recipient. In addition to these general obligations, in matters of maritime travel, there is an ongoing responsibility to ensure the nautical and commercial fitness of the ship. The transporter also has specific obligations: taking delivery, loading, handling, docking, safeguarding, caring for and unloading. From the receipt to the delivery of the merchandise, the transporter's responsibility is based on a presumption of responsibility, but it may exonerate itself from this by proving one of the cases of exemption provided for by the text. The loader may still refute the transporter's defense, however, by proving that the damages are imputable to it or by establishing that this damage is not the result of a case of exemption.

– Most of the French law is contained within the transport code. This text concerns all types of maritime transport contracts, and is applicable from the time the merchandise is taken in hand until its delivery.

As an instrument of governance, the bill of lading is the principal supporting document of a maritime transport contract. The fruit of longstanding historic tradition, a bill of lading can be made to the order of or to the bearer. Issued at the request of the loader, who is no longer required to sign it, it is filled out by him/her, by the transporter or by the transporter's representative based on the information provided by the loader, who is then responsible for the accuracy of the indications relative to merchandise, with any inaccuracies engaging the loader's responsibility to the transporter. The bill of lading is issued in at least two original copies; one for the loader and the other for the master. The law specifies the information that this document must contain: proper names to identify the parties, merchandise to be transported, facts about the voyage to be undertaken and freight cost to be paid. It must also indicate adequate brand information to identify the merchandise; the quantity of this merchandise (in numbers of packages or in weight) according to the information given by the loader; and finally the apparent state and storage of the merchandise.

2.3.2.1. *Obligations of maritime carriers*

Aside from its central obligation "to delivery a given type and amount of merchandise from one port to another", conventions and the law impose a series of obligations on the transporter. According to the 1924 convention, "the transporter [is] required before and at the start of the voyage to exercise due diligence in order to a) ensure that the ship is seaworthy; b) fit out, equip, and stock the ship adequately; and c) ensure the good condition of [all parts of] the ship where merchandise is loaded, for its reception, transport, and conservation". Unloading operations, which are the responsibility of the transporter, must take place in conditions analogous to those of loading.

Stowage on deck, which consists of arranging merchandise on the deck of a ship rather than in the hold, is a risky technical and commercial operation that has given rise to debate. From this point of view, positive law distinguishes between regular on-deck loading – that is carried out in accordance with legal specifications – and irregular on-deck loading, which does not comply with these specifications. According to the La Haye-Visby rules, stowage on deck is regular if it has been declared thus on the freight bill with the agreement of the loader and then loaded in the agreed-upon way; if it fulfills this double condition, it will be exempt from its field of application. If this condition is not fulfilled, the transporter will be at fault; depending on the circumstances proper to each case. Similarly to those of French law, the Hamburg rules do not exclude on-deck transport from their field of application, specifying that it will be considered regular if the loader has given its consent or if this mode of transport is required by regulations or if it is carried out in accordance with the customs of the trade concerned. According to this text, on-deck transport that does not meet these conditions may constitute an inexcusable transgression on the part of the transporter, thus depriving it of the right to limit the consequences of its responsibility. The Rotterdam rules set forth hypothetical cases in which on-deck transport is permitted and non-transgressive; if the deck is required by the law; if it is in compliance with the customs, usages and practices of the trade concerned; if it is in compliance with the transport contract (that is if it is undertaken with the consent of the loader); and, finally, if the loading of containers or vehicles takes place on decks that are specifically equipped to transport them. In these cases of regular on-deck transport, the transporter's responsibility will be engaged in accordance with the terms of the convention, except in the event of loss, damage or delay (resulting) from the specific risks involved in this type of transport. If the on-deck transport does not fulfill the conditions of the text, the transporter cannot claim exemption from responsibility or invoke limitation of responsibility if the transport was undertaken even though the transporter had expressly agreed with the loader that the merchandise would be transported in the hold.

2.3.2.2. *Obligations of loaders*

Loaders are required, like all beneficiaries of merchandise, to present this merchandise in accordance with the conditions of time and place specified by the contract. If this is not the case, the loader will owe the transporter a compensatory sum corresponding to the damage sustained, within the limit of the sum of the freight cost. Likewise, the costs of shipment and freight due in order to complete transport of the merchandise are the responsibility of the loader provided that the interruption of the voyage is not due to the fault of the transporter; otherwise, these costs are its responsibility. In both scenarios, the transporter keeps the freight cost specified for the whole voyage.

The loader must also compensate the transporter for damages caused to the ship or to other merchandise due to its error or by the defects of its own merchandise. Finally, it must take delivery of this merchandise; barring a claim on the merchandise or in the event of contestation relative to delivery or to payment of freight costs, the captain may, by legal authority, have the merchandise sold in order to pay freight costs and order any surplus to be stored.

The loader must pay the costs of transport (or freight). Though the freight fees are in principle set by the parties according to the weight or volume of the merchandise, it is sometimes affected by various additional costs and fees (loading and unloading fees, customs duties, etc.). It may be agreed upon that freight costs are payable in advance or upon arrival at the destination. In the latter case, the receiving party is also the debtor if it accepts the delivery of the merchandise; on the contrary, if it refuses the delivery, the freight costs will be payable by the loader.

If the freight costs remain due or liable for taxes, for merchandise thrown overboard into the sea for the common safety, these costs will no longer be payable for merchandise lost due to the hazards of ship transport or following the transporter's failure to fulfill its obligation to keep the ship fit to sail or if the loss is due to a failure to fulfill its obligations relative to the merchandise.

In order to protect itself, the transporter may insert into the marine bill of lading a "freight cost acquired in any case" clause, which will enable it to collect the freight cost despite the loss of the merchandise for any reason, "whether perils of the sea or otherwise". If payment is not made, it will still be protected by the law granting it the right to retain the merchandise on board.

2.3.2.3. *Responsibilities of maritime carriers*

For losses and damages sustained by merchandise with which it has been entrusted, the Brussels Convention and French law specify the transporter's responsibility by full public right in its conditions and effects. It is thus responsible for losses or damage sustained by the merchandise from the time it is taken in hand to the time it is delivered, unless it can be proven that these losses or damages were caused by a limited number of specified facts. The Hamburg rules seem to establish a presumption of fault, but not responsibility, in the sense that the transporter is deemed responsible unless it can prove that all measures were taken that could reasonably be expected to avoid the incident and its consequences; that is that no damage-causing transgression was committed by it. The Rotterdam rules are similar; they increase the transporter's responsibility by pronouncing its responsibility for all damages sustained by the merchandise unless it can prove that neither its own fault, nor that of its employees, caused or contributed to the damage. The transporter is also exonerated if it can prove that the damage was caused by an excepting event, the list of which is similar to that put forth by the 1924 convention.

Justified by the idea that the incidence of risks inherent to travel at sea must be shared among its participants, the mechanism to limit the responsibility of the transporter is seen as compensation for the strict liability weighing on it. In this, derogating from ordinary law which requires the party responsible for damage to pay the reparations in full, whereas the limitation enables a transporter known to be responsible for damages to pay reparations only up to a threshold determined by referring to the 1924 convention. However, the transporter is not entitled to this mechanism if the losses or damages result from its intentional or inexcusable transgression. The presumption of responsibility imposed on maritime transporters by the law is not indisputable, as attested to by the possibility available to them of exonerating themselves by proving that a given damage sustained by merchandise has been caused by an exceptional incident with responsibility attributable to, among other causes, the conduct of the loader, the operation of the ship or an outside event with characteristics of force majeure. These exceptional cases, which make it possible for the transporter to exonerate itself more easily than a debtor with a contractual obligation under ordinary law, are identical in substance and form in the 1924 convention and the law, though their formal presentations differ in the two texts.

2.3.3. *Maritime insurance*

Because a ship and its cargo must be insured against any damages they may sustain or cause, maritime insurance is intended to manage damages that arise as part of a maritime operation. Insurance has always been important at sea; since the high Middle Ages, shipowners have been able to protect themselves against the perils of the sea via a "Bottomry loan", in which they borrowed a sum corresponding to the value of the ship and the merchandise being transported. In the event of the ship's safe return to port, the borrower was obliged to repay this sum increased by a premium agreed upon as the price of the risks incurred. If the ship was lost, the shipowner's repayment obligation was rendered void.

Insurance, as a method of collectively distributing the risks of accidents at sea, is the condition *sine qua non* for the efficient governance of maritime commerce; the enormity of the capital involved has made recourse to maritime insurance indispensable. Article L.171-6 of the insurance code classifies "maritime vehicles as well as the risks of responsibility pertaining thereto" and "transported merchandise" among the "major risks" requiring specific regulations.

These contracts are random in the sense that the benefit or loss that may result from them depends on an uncertain event, and maritime insurance is characterized by the maritime nature of the risk being considered. In order to be insurable, rights must be subject to the risks of maritime navigation; this rule results from the fact that maritime insurance "is intended to underwrite the risks involved in a maritime operation" (C. assur., art. L.171-1). Since the *Ordonnance de Colbert* of 1681, the risks of the sea have been referred to as the "fortunes of the sea" (from *fors fortuna:* (un)favorable outcome, risk), which evokes the fortuitous outcome (*fortuitus* arises from *fors*). This expression, which shows the random character of maritime insurance, encompasses all of the dangers of navigation that may strike a ship and its cargo during a maritime operation. As a marker of maritime insurance, the "fortunes of the sea" are widely understood to apply to any risk that may arise during maritime navigation, whatever its cause. For this reason, it is not only incidents at sea themselves that are classified as risks of the sea, but also a number of aftereffects directly caused by these incidents.

Though insurable risks have been elaborated on to a great extent, unexpected hazards constitute an impassable limit to this elaboration. The "fortunes of the sea" will always exist when a contract is concluded; in the absence of a hazard, the contract will be void, as it is without purpose.

Modern maritime insurance contracts are regulated by the insurance code. The law determines a complete legal corpus which first sets out the general rules common to various types of insurance, and then distinguishes the three categories of maritime insurance: hull insurance (for ships); freight insurance (for merchandise); and liability insurance, which enables shipowners to protect themselves against the risk of liabilities not covered by hull insurance.

Unlike land insurance, maritime insurance law leaves a great deal of room for the expression of contractual freedom; with the exception of those specified by the law, legal regulations can be set aside by the parties. Virtually, all contracts are concluded using boilerplate models, which are regularly updated to take technological and legal developments into account.

There is a French maritime insurance policy for hull insurance (last updated January 1, 2001) and several French maritime insurance policies for cargo insurance, including protection against "all risks" and protection against "specific risks barring major events" (last updated July 1, 2009), which can be taken out for a single expedition or be the subject of a subscription for successive expeditions. French insurers also offer liability insurance policies for shipowners (December 20, 1990) and maritime transporters (December 20, 1972); there are also special policies against risks of war or the equivalent.

2.3.3.1. *General obligations of insurers*

The principal obligation of an insurer is the payment of insurance benefits if a peril of the sea occurs under the conditions specified in the contract. Policyholders may be compensated for all harmful consequences of incidents covered by the policy. With the exception of physical injury, the damage sustained by the policyholder can be material loss or damage, commercial damage, an incurred expenditure or third-party action taken against the insured party. In the case of insurance covering fire, for example, all injurious consequences of a fire should in principle be borne by the insurer. This is the characteristic service provided by an insurance contract, which does not mean, however, that every injurious consequence of an incident covered by a policy will necessarily be guaranteed.

In addition to excluding certain causes of harmful incidents proper to each policy, insurers do not cover damages resulting from the defects of the ship itself or of the merchandise insured. Indeed, insuring the defects and flaws presented by the insured object would mean denying the random character of the contract, as their existence would make a disaster highly probable.

Likewise, covered risks remain covered even if the insured party is at fault, unless the insurer determines that the damage is due to a lack of reasonable care on the part of the policyholder to protect its assets from the risks incurred; coverage of an intentional or inexcusable fault on the part of the policyholder is prohibited by the provisions of the insurance code.

2.3.3.2. *Hull insurance*

Hull insurance covers the ship and all its equipment. It covers the ship while it is being constructed; its freight, and the maritime operation. Here, the term "ship" means the hull and its locomotor system as well as all the accessories and attachments necessary for its use, and the costs of fitting out and supplying the ship.

Hull insurance covers damage liable to be sustained by the ship, up to and including its total loss. Accidental damage is covered, as is some deliberate damage and damage resulting from a decision made by public authorities with the intention of preventing or reducing pollution, if the origins of the risk of this pollution lie in a covered incident. If, after a covered collision, a ship transporting toxic products is scuppered in order to avoid a polluting incident, the coverage will hold.

Hull insurance covers the ship's responsibility for a collision. It also covers the accessories and equipment on board the ship at the time of the collision, and third-party actions against the ship for damages caused by its machinery, anchors and chains or by any small boats attached to it. However, hull insurance covers only material damage caused to third parties, excluding claims for physical injury. In terms of its amount, the liability coverage provided by the hull insurer (damages, losses and claims made against the insured hull) is limited to a sum equal to the value of the ship, called the "approved value". A typical policy specifies the limit of the insurer's responsibility at an amount equal to two times the approved value. Coverage is valid on a "per incident" basis; if a ship is involved in two successive collisions, the insurance may pay up to two times the approved value for each incident.

Under the terms of article L.172-16 of the insurance code, the insurer does not cover incidents arising from civil or foreign war, piracy, or riots, among other causes, or those due to the effects of atomic explosion or radiation.

The obligations of the policyholder are determined by article L.172-9 of the insurance code. The insured party must: (1) "Pay the premium and costs at the agreed-upon time and place"; (2) "Take reasonable care in all matters pertaining to the ship or merchandise"; (3) "Declare precisely, at the time of conclusion of the insurance contract, all circumstances known to it that may increase the risk taken by the insurer"; and (4) "Declare to the insurer, insofar as is known to it, any increased risks arising during the course of the contract".

2.3.3.3. Cargo insurance

Cargo insurance covers damage and losses pertaining to merchandise. Though this insurance connects the insurer to the loader (the owner of the merchandise), the subscriber and the beneficiary of the insurance policy, is in reality at the core of the process. In practice, the beneficiary of the policy is not designated by name, with coverage being contracted "on behalf of the party to whom it will belong". In this case, this contractual detail is considered to constitute both insurance for the benefit of the policy subscriber and a stipulation for others for the benefit of the beneficiary of the said clause (C. assur., art. L.171-4). Consequently, the coverage may be invoked by the subscriber or by the owner of the merchandise at the time of damage (the recipient). This detail of cargo insurance thus makes it possible for the coverage to follow the merchandise as it changes hands several times. This change of beneficiary occurs frequently when merchandise is sold along the way.

Two levels of coverage coexist in cargo insurance: protection against "all risks" and protection against "free of particular average (FPA)". While the first level very broad category covers all insurable perils of the sea except for those that are expressly excluded, the second level, which is more limited, covers only the specific damages listed in the policy. In FPA coverage, the risks covered include all major incidents that may arise during the course of maritime, land, aerial or river transport of merchandise: shipwreck, capsizing, running aground, collision, watre ingress requiring the ship to enter a port of refuge; falling of packages, accident of land transport vehicle, flood, volcanic eruption, fire, explosion, and aircraft crash, among others. In addition to the extent of each type of coverage, there is a significant difference between the two in terms of proof. While it falls upon the "all risks" insurer to prove that a case of damage is excluded from coverage in order to be freed from its obligation, it is the responsibility of the "FPA" insurer to prove that the damage is the result of a covered risk.

Depending on the policy type, "coverage begins at the time the cargo [...] is moved in the warehouse at the extreme starting point of the insured voyage, to be immediately loaded onto the transport vehicle". It is completed at the time delivery is taken of the cargo by the policyholder or the subscriber. The time-management involved in cargo insurance goes beyond the domain of maritime risks; thus, in a multimodal transport operation, this insurance will cover all land transport operations preceding or following maritime transport, with a limit of 60 days calculated from the completion of the unloading of the cargo from the last ship.

Cargo insurance is principally damage insurance. Coverage, within the limit of the approved value (that is the price of sending the cargo to its destination), includes damages and losses sustained by merchandise, including those caused during loading or unloading carried out by the policyholder or beneficiary of the insurance policy. Subject to the holding by the transporting vessel of a safety management certificate, costs reasonably incurred in order to preserve insured cargo from a covered incident of damage or to limit the consequences of this damage, contribution to joint damage, or remuneration for assistance, will be covered up to their full amount and proportionally to the value insured.

The 2009 insurance policy contains a number of general exclusions similar to those given by hull insurance. Notably, exclusion for intentional or inexcusable fault of the insured policy is more extensive in cargo insurance, since it applies to the fault of the insured party or the fault of the beneficiary of the policy, as well as to faults committed by their employees, representatives or assignees. Except for the latter fault, excluded risks can still be covered if an additional premium is paid.

The obligations of cargo insurance policyholders are nearly identical to those of hull insurance policyholders. As part of its duty of honesty at the time the contract is concluded, the insured party must first declare to the insurer all of the circumstances that may increase the risk taken by the insurer in covering the operation. The insured party must also declare any increased risks arising during the course of the contract. The insured party, like any beneficiary of the insurance, must take reasonable care with regard to everything pertaining to the merchandise; it must, therefore, take all possible protective measures to prevent disaster or to limit the harmful consequences of such a disaster. Finally, the insured party must take all possible measures to protect the rights and recourse of the insurer against the transporter or any other responsible party.

2.3.3.4. Protection and indemnity (P&I) clubs

Since 1855, Protection and Indemnity Clubs (P&I Clubs) have existed as groups of shipowners covering financial and liability risks that are not covered by hull insurance, or which insurance companies refuse to cover. Operating in mutual benefit mode, in which the sum of the members' annual dues marks the limit of the total coverage by the club, P&I clubs have historically played a role in human solidarity at sea. Today, these clubs cover around 90% of maritime risks with civil liability. Unlike hull insurance, where the approved value, unless otherwise specified, marks the limit of the insurer's engagement, the protection given by P&I Clubs is unlimited.

The risks covered by P&I Clubs are specifically determined by each club. The most significant coverage lies in protecting shipowners in their relations with their co-contractors. This would be the case for loaders in the event of damages to merchandise; for passengers in the event of injury or fatal accident; and for seamen working on board ship in the event of fatal accidents caused to third parties. The clubs cover risks that would not be covered by hull insurance (responsibility for collision, remuneration for assistance, contribution to joint damages and now responsibility in the event of pollution). The same is true for certain financial responsibilities such as payments made by shipowners in the event of the death, injury or illness of sailors; costs of destruction or raising of a shipwreck; or fines levied against the shipowner in the event of a breach of customs regulations or an infraction in matters of immigration.

2.4. Bibliography

[COR 11] CORNU G. (ed.), *Vocabulaire juridique*, 10th ed., PUF, Paris, 2011.

[REM 98] RÉMOND-GOUILLOUD M., "Evénements de mer et responsabilité", *ADMO*, vol. 16, 1998.

[SCH 81] SCHADÉE D., "La mer comme mère du droit", *Etudes offertes à René Rodière*, Dalloz, Paris, 1981.

[VIA 97] VIALARD A., *Droit maritime*, PUF, Paris, 1997.

[VIG 87] VIGARIÉ A., *Echanges et transports internationaux*, Mémentos de géographie, Sirey, Paris, 1987.

Marine Pollution:
Introduction to International Law
on Pollution Caused by Ships

3.1. Introduction

Marine pollution, whatever its source, has long been an ongoing concern for governments, the public and environmental advocates. Yet, for nearly 50 years, this pollution has only continued, growing worse and more varied, and we may wonder what the law is doing to contain this problem efficiently.

Because the sea is an international space, it has naturally fallen – at first, at least, and mainly – to international law to address the issue.

This sector of law has been built bit-by-bit, made up of international conventions that have often multiplied, sometimes more than once, in reaction to a specific event so that this event does not occur again.

However, a real innovation was introduced by the United Nations Convention on the Law of the Sea of December 10, 1982, called the Montego Bay Convention[1] (and generally referred to as UNCLOS). This was not the first time that the formally organized international community had taken an interest in marine pollution, and the United Nations Geneva Convention on the High Seas of April 4, 1958 included several rules

---

Chapter written by Véronique LABROT.
1 Effective date December 16, 1994.

concerning pollution by hydrocarbons. These do not compare with the ambition of UNCLOS, however, which devotes Part XII, or articles 192 to 237, to the "Protection and preservation of the marine environment", specifying in its preamble "that it is desirable to establish, by means of this convention, and duly taking into account the sovereignty of all States, a legal system for the seas and oceans that [...] advances [...] the protection and preservation of the marine environment".

Article 194 of UNCLOS targets all forms of marine pollution[2], and article 192 makes it "the business of governments[3]", flag States[4] and coastal[5] or port States[6]. Measures taken to address these questions may be taken separately[7] or jointly[8].

The committed stance of international law, of part XII of UNCLOS, arising from the framework agreement included in the United Nations Convention on the Law of the Sea, has imposed on States compliance, in particular, with "generally accepted regulations and standards, established via the intermediary of a competent international organization...[9]". These rules, which are not specified by UNCLOS and most of which do not come from the UNCLOS Convention itself, are thus mainly preexisting; that is very specifically, the operational or accidental pollution of the seas by ships and hydrocarbons[10] as laid out by the conventions of the International Maritime Organization (IMO)[11], the only United Nations agency specializing in

---

2 See part XII of UNCLOS which is more or less precise depending on the type of pollution.
3 According to article 192, "States have an obligation to protect and preserve the marine environment".
4 Article 217 of UNCLOS.
5 Article 220 of UNCLOS.
6 Article 218 of UNCLOS.
7 Including in unilateralism, such as the position of the United States following the 1989 Exxon Valdez catastrophe in Alaska, with regard to the development of the Oil Pollution Act of August 18, 1990 – see [REM 91].
8 Article 194.1 of UNCLOS.
9 Article 211.2 of UNCLOS.
10 Though the principal form of maritime pollution is telluric pollution, which is very scantily regulated, and though pollution by ships by hydrocarbons remains a lesser problem, media coverage of black tides has led international law to develop mostly with regard to marine pollution by ships; therefore, this will be the only type of pollution discussed here.
11 Other regional international organizations such as the ILO have of course intervened, usually in ways compatible with IMO conventions; see European Union policy on the subject, for example.

matters of global navigation security. Hence, it is a case of relying on international law on marine pollution by ships, since the actions of this law are aimed as much at the prevention of pollution (section 3.2) as at intervention in the event of an accident (section 3.3), and at the repair of damage caused by pollution (section 3.4).

3.2. Preventing pollution by ships

It is clear that, at a time when there are increasingly serious and sometimes irreversible damage-causing events, the prevention of pollution is crucial.

Though the conventions pertaining to pollution provide regulations that are often considered adequate to protect the oceans in their entirety, there is another environmental reality taken into account by international texts: the existence in this already fragile aquatic space of even more vulnerable marine zones which deserve in various ways greater protection than that applicable to the oceans as a whole. It is also important to recognize that in addition to the political zoning of the sea provided for by UNCLOS there is a superimposed system of ecological zoning (section 3.2.1). It falls to the law, then, to design maritime routes that are both safe (section 3.2.2) and clean (section 3.2.3).

3.2.1. *Spatial preconditions: acknowledgment of protected maritime zones*

The International Law of the Sea, as now largely contained in the United Nations Convention on the Law of the Sea, has developed an approach involving a shared ocean divided into an ever-increasing number of maritime zones over which coastal States exercise more or less authority; an authority which diminished as distance from the coast increases. Thus, starting from land territory and moving out to sea, we reach inland waters, then territorial waters, then the contiguous zone, and then the exclusive economic zone containing all or part of the continental shelf, and then the high seas and finally, beneath the high seas, the International Zone. To these "classic" divisions, other specific zones have been added, such as straits, archipelagic waters, enclosed or semi-enclosed seas, ice-covered zones, etc., not to mention transoceanic canals not envisaged by UNCLOS. There is also a system of zoning superimposed on this political/administrative legal marine zoning that is equally diversified but not necessarily compatible with UNCLOS zoning; this is ecological zoning, and it is this that is being referred to when "protected marine areas" are mentioned.

Indeed, it is a current trend to establish "Marine Protected Areas" (MPAs)[12] as attested to by the number of States that have committed to protecting them[13], in the same way as specified in a sometimes limited manner by UNCLOS[14] and as currently being put in motion by the European Union via its "Sea Natura 2000" initiative. The European Natura 2000 initiative, intended to create a network of protected sites throughout the territories of member States (and subsequently the European Community) via application of the so-called "Oiseaux" (or "Birds") directive (directive 79/400/EEC of April 2, 1979, subsequently replaced by directive 1009/147/EC of November 30, 2009) and, especially, by the "Habitats" directive (directive 92/43/EEC on the conservation of the natural habitats of wild fauna and flora of May 21, 1992), is an Initiative the European Union (following decision by the European Union Court of Justice C26/04 of October 20, 2005), of which asked from now on for an application at sea, in marine areas which would be quite suitable for the application of the Oiseaux and Habitats directives. This was made more concrete by the development by the European Commission in May 2007 of "Guidelines for

---

12 The creation of which on the high seas has very recently been quite specifically defended by the Global Ocean Commission's report entitled "From Decline to Recovery – A Rescue Package for the Global Ocean", June 24, 2014, available at the Commission's Website: http://issuu.com/missionocean/docs/goc_full_report/0.

13 For example, the recent classification by Nauru, Pitcairn and Palau of their exclusive economic zone as a preserve, or even sanctuary, enabling the creation of immense marine reserves (603,678 km² for Palau, for example, which is larger than the land area of mainland France). See also the American proposal to classify the Pacific Ocean as a preserve, and the Kiribati proposition in the June 17, 2014 edition of L'Express. See also the French decisions after the Grenelle of the Sea to create a blue framework or corridor via the setting up of a network of MPAs covering 10% of the French EEZ by 2012 and 20% of this French maritime space by 2020.

14 See article 194.5, which evokes the measures needing to be taken "to protect and preserve rare or delicate ecosystems as well as the habitat of species and other marine organisms in decline, threatened, or in danger of extinction"; see also article 234 on "ice-covered zones" or part IX of UNCLOS on enclosed and semi-enclosed seas, completed by the complicated "regional seas" program of the UN Environment Program (UNEP), instituted in the mid-1970s.

the establishment of a Natura 2000 network in the marine environment",
addressed to Member States obligated to comply with it in order to identify
these Natura 2000 zones in their maritime spaces. This was the case, for
example, for the Lavezzi Islands, the Strait of Bonifaco and the Bay of
Morlaix in France. The Natura 2000 system is legally provided for in France
by articles L414-1 to 414-7 of the environmental code. Maritime sites are
managed by the current Protected Marine Areas Agency, created by the law
of April 14, 2006 and completed by the "National strategy for the creation of
PMAs – a doctrinal note for metropolitan waters" of November 20, 2007.

All of this has contributed to the implementation of ecological zoning,
often in the form of networks of sites to be protected, in which the
international level appears more like a degree of motivation to be determined
at the regional or national level of these PMAs, sometimes like a level of
decision-making by an international organization, but on request of the
States[15]. Such is the notable case in matters of marine pollution by ships –
though sometimes insufficient – of the IMO.

Since the 1970s, the IMO has been defining what they call "protected
maritime zones" (PMZs), as "intertidal or infratidal zones with the waters
covering them, their flora and fauna, and their historic and cultural heritage,
which have been categorized with a view to protecting all of part of the
environment they compose".

This very broad definition enables a multitude of forms of protection of
the aforementioned zones (marine preserves, natural parks, marine parks,
sanctuaries, protected sites, etc.):

– spontaneous national protection, first in zones under the sovereign
control of coastal States[16];

– incentives and then international certifications followed by national
modes of protection[17], forms of regional protection imposed on States[18] or

---

15 One exception may be seen, for example, in the decision made directly in 1994 by the
International Whaling Commission to classify the Antarctic as a sanctuary for cetaceans.
16 See article L334-1 and following of the French environmental code on "natural marine
parks" for example. It should be noted that the majority of PMZs are found near coasts; that is
in inland seas and territorial waters.
17 See "UNESCO natural world heritage sites" and their protection via their designation in
France, for example, as preserves or registered sites, among other existing provisions.
18 See the Natura 2000 sea network for the aforementioned European Union Member States.

forms of international protection determined, at the request of coastal States, by a system of zoning established by the IMO in particular (if it is a matter of protection against pollution by ships) when it may also have effects reaching beyond territorial waters.

These zones created by the IMO include particularly sensitive sea areas[19] (PSSAs[20]) and "special zones" (SZs) MARPOL[21].

These two principal types of PMZ have quite comparable definitions, though in detail they do not require the same ecological, scientific, economic, cultural or other criteria[22].

The IMO defines a PSSA as a "maritime zone which, due to its recognized ecological, socio-economic, or scientific importance, should be the subject of special protection via measures taken by the Organization [that is, the IMO] and which may be vulnerable to damage caused by maritime activities"; and an SZ MARPOL is a "maritime zone which, for acknowledge technical reasons having to do with its oceanographic and ecological situation as well as the specific character of its traffic, calls for the adoption of particular required methods in order to prevent marine pollution by [hydrocarbons, chemical products, wastewater, etc., according to the appendices]".

This situation results in a sort of "striated" sea, in which can be seen, superimposed on the various areas of authority of coastal nations specified by UNCLOS[23], "protected marine areas" whose legal systems will be more drastic in matters of environmental protection than for the rest of the seas and oceans, which are more generally protected. Roads through these areas must, as in the rest of the oceans, be safe and clean. To ensure this, since most international conventions having to do with preventing pollution by ships arise from the IMO, the latter can use only protective "tools" in compliance with its "social objective", which is to ensure the safety of

---

19 In 2011, the IMO agreed to classify the Strait of Bonifacio as such, an international strait separating Corsica and Sardinia and containing particularly rich biodiversity representative of a future international marine park.
20 For Particularly Sensitive Sea Area.
21 Or "emission zones" for appendix VI of MARPOL – see *infra*.
22 See, for example, the IMO circular, MEPC 1/Circ.778 of 01-26-2012.
23 Territorial waters, contiguous zone, exclusive economic zone, etc.

transport and not, notably, the protection of species[24], which then occurs only indirectly through the application of IMO instruments. These instruments are intended to use the famous Donaldson report title of 1994, ensure "Safer Ships, Cleaner Seas"; that is to act on the traffic of ships on one hand, and on their construction on the other hand. While this has long served to "protect ships from the sea", today it also helps to "protect the sea from ships" to use the expressions coined by Professor Martine Rémond-Gouilloud [REM 93].

3.2.2. *Safe routes: the organization of maritime traffic in question*

The primary thing that may prevent pollution is undoubtedly, alongside the use of "clean" ships, ensuring that maritime routes are safe enough to avoid the causes of major pollution, collisions, by providing for vessel traffic service (VTS), which takes into account all pollution-causing accident risks in order to reduce the occurrence and consequences of these. These VTS measures are organized principally under the aegis of the Convention on international regulations for preventing collisions at sea, or COLREG[25], as well as in Chapter V of the SOLAS convention (convention on the safeguarding of ships, hygiene and habitability on board and the fight against pollution)[26], not to mention the training of seafarers provided for by the Convention on standards of training, certification and watchkeeping, or

---

24 See, however, the criterion for the protection of resources in the definition of special zones, as an objective added to article 211-6 of UNCLOS, which refers to it. See also, in the "regional seas" program of UNEP, certain conventions combined with protocols pertaining to the protection of biodiversity via the establishment of "specially protected areas and specially protected areas of Mediterranean importance" (SPA and SPAMI), with the Pelagos sanctuary standing as the primary example of a SPAMI.

25 Or Collision Regulations, the 1972 IMO convention made effective on July 15, 1977, amended several times since.

26 Convention which also provides, in rule 8-1, for the implementation of compulsory reporting (CR), which captains must do – normally at the request of a flag state, but most often at the request of a coastal state – upon entering an at-risk zone such as, in France, before entry into the rail d'Ouessant (see *infra*): identity of ship, port of departure, destination port, cargo contents, etc. (SURNAV). These mandatory reports go along with a number of provisions specified by this "code of the sea", especially for ships transporting polluting substances. These CRs are now required before entry into any port of the European Union, following directive 2002/57/CE of June 27, 2002 (the Erika II packet) modified in 2009 (Erika III packet), which notably created the traffic and information monitoring system (Safe Sea Net system), also modified in 2009, and addressing the question of ports of refuge.

STCW convention[27], not seen here[28]. COLREG contains mainly regulations pertaining to steering and sailing, and is best known for its rule 10 relative to provisions for the traffic separation scheme (TSS)[29], which provides, in difficult passes, for the setup of "sea highways" with two or more lanes, each with a direction – climbing or descending – and divided by a separation zone in which traffic, except for perpendicular crossings, is prohibited; these are often accompanied by "caution zones". These TSSs cannot be put in place by coastal States without the prior consent of the IMO if the TSS is partially or wholly outside of territorial waters, or only after declaration if it is fully within territorial waters or a strait less than 24 nautical miles wide.

In order to protect areas that are vulnerable to pollution by ships, particularly following accidents[30], States often turn to the setup of these types of TSSs. However, other measures are taken as well, in particular to protect a PSSA. All these measures are relative to the conditions of ship traffic presenting a risk of pollution.

The IMO can allow the setting up of "deep water routes" that are sometimes recommended, but may also be imposed on deep-draught ships for areas where hydrographic studies are non-existent or inadequate. In this case, this lack of knowledge may cause doubt with regard to the depth of the sea-bottom and thus to the ability of certain classes of ships to pass through a given zone with enough molded depth[31]. This is also the case for areas where hydrographic studies specify the depth of the sea-bottom and the existence of submerged objects.

---

27 In its modified version and to which several International Labor Organization (ILO) conventions have been added in the same area.

28 We refer readers to Chapter 2 of this book.

29 Still called "rails", the best known of which in France is off the coast of Finistère, the rail d'Ouessant is affected by the TSS of Pas-de-Calais and is currently in the process of being modified.

30 See, for example, the implementation of the TSS as a mode of protection of the marine environment, provided for by articles 22 (territorial waters), 41 (straits), 53-4 and following (archipelagic waters), 211, etc., of UNCLOS. See the implementation of the TSS in the Bosphorous strait, which is extremely narrow but has been very highly frequented, particularly by supertankers, following a maritime accident in 1979 suffered by the L'Independenta, which caused a fire in Istanbul, a coastal city. See also the TSSs in the strait of Singapore.

31 See, for example, the creation of a deep-water route on the outskirts of King Abdullah port on the northern coast of the Red Sea (Saudi Arabia) or the modification of the existing deep-water route in the Pas-de-Calais TSS.

Conversely, the IMO may establish "avoidance zones", meaning, according to this organization, "a zone located within predetermined limits, within which navigation is particularly dangerous, or within which it is particularly important to avoid accidents and which should be avoided by all ships or certain classes of ships[32]".

Following the disasters suffered by the *Sea Empress* in 1996, the *Erika* in 1999 and the *Prestige* in 2002, France, as well as other coastal States affected by these catastrophes – Ireland, Spain, the United Kingdom, Portugal, etc. – requested that an entire part of the North Atlantic economic zone, from southern Portugal to the Celtic Sea (excluding the North sea) as a PSSA, to be protected via the establishment of an area to be avoided (ATBA) for supertankers. The IMO refused to grant this classification, however, except for non-double-hulled tankers of more than 600 gross registered tons (grt) transporting heavy oil. Though it judged ATBA classification to be overly excessive, the IMO acknowledged the existence of a PSSA in order to attract crews' attention to the vulnerable nature of the marine environment being crossed, and allowed the States concerned to require other ships and double-hull tankers to complete a CR 48 h before entering the area, but did not allow the requirement of compulsory pilotage, which the IMO also refused for the Strait of Bonifacio, preferring simply to designate it as an area where "deep-sea pilotage [is] strongly recommended[33]".

The institution of and compliance with all these measures, whether compulsory or not, are ensured by the coastal State (or States, if there are several) via the implementation of [34] VTSs, which range from the simple broadcasting of messages to ships (meteorology, status of sea traffic, etc.) to the use of more extensive services such as TSS.

---

32 For example: the creation of an ATBA in the Atlantic Ocean off the coast of Ghana; the establishment of a compulsory anchoring-prohibited zone for all vessels and an ATBA for ships of more than 300 grt; related protective measures for the preservation of the Banc de Saba PSSA; and an ATBA on the Australian Great Barrier Reef, greater pressure in the enforcement of which was requested of the IMO in 2014.

33 See, for example, the motion of the Corsica Assembly of January 27-28, 2011 no. 2011/ E1/002.

34 These include *centres régionaux opérationnels de surveillance et de sauvetage* (CROSS) [regional operational search and rescue centers] in France, including CROSS Etel A and CROSS Corsen, which monitor traffic in the rail d'Ouessant in the open seas off Brest, but may also be simple buoys delineating a recommended route or pilotage service.

The existence of a "code of the sea" of this type would not be enough if ship design was not also part of the picture, as safer ships enable cleaner seas.

3.2.3. *Clean routes: design and management of the ships in question*

If it is not enough for navigation rules to be established or even respected in order to protect the sea from ships and ensure maritime safety, it is because these ships must themselves be designed as "clean ships". In this, design and management are crucial. Solid ships must be constructed, and the 1966 IMO Load Lines convention, for example, is an important part of this. These ships must be as clean in terms of both construction and procedures (particularly having to do with waste disposal) as possible. For this reason, most ship-caused pollution is addressed in IMO texts, of which only the current MARPOL convention is legally binding at the moment (section 3.2.3.1), while "newer" forms of pollution are still addressed through existing texts, most of which are not yet binding, at least at the international level (section 3.2.3.2).

3.2.3.1. *From OILPOL 1954 to MARPOL 1973–1978: principal binding laws*

Some regulations existed before the 2nd World War, mainly at the national level; these had to do mostly with operational hydrocarbon pollution in ports. However, it was not until the post-war years that the international community, faced with the challenge of developing maritime transport for mineral resources such as hydrocarbons on ships that were constantly growing larger and larger, and thus more dangerous for the environment in the event of collision or beaching[35], finalized the first convention concerning "marine prevention of pollution by hydrocarbons" in 1954, known by the acronym OILPOL, for Oil Pollution. It was placed under the responsibility of the IM[C]O[36] as soon as the convention establishing this new international organization went into effect. Despite the innovations introduced by OILPOL, it quickly proved inadequate, and, especially after the disaster suffered off the Isles of Scilly by the Liberian oil tanker *Torrey Canyon*, the

---

35 See the construction of so-called "pre-MARPOL" ships.
36 The IMO was first created in 1958 as the Inter-Governmental Maritime Consultative Organization (IMCO).

IMO attacked the problem of hydrocarbon pollution with new vigor. A new, broader convention was developed "for the prevention of marine pollution by ships", or MARPOL, in 1973. Because its entry into force proved a lengthy process, it was modified to speed up its applicability in 1978.

Today, though some rules for the prevention of marine pollution can be found in other IMO conventions, notably within various chapters of the 1974 SOLAS convention[37], it is the MARPOL (for Marine Pollution) convention that includes, to quote the aforementioned article 211 of UNCLOS, "generally accepted international regulations and standards" applicable to pollution by ships.

Unprecedented in its composition, this MARPOL convention is formed of a framework convention containing its general conditions of application[38], accompanied by two protocols; one on the settlement of disputes between signatory States, and the other on the sending of reports pertaining to events causing or with the ability to result in waste composed of harmful substances. It essentially defines the technical regulations thus imposed on signatory States in a number of appendices which currently stand at six and which are regularly amended and concern various sources of marine pollution by ships:

– appendix I: pollution by hydrocarbons;

– appendix II: pollution by chemical products transported loose in bulk;

– appendix III: pollution by chemical products transported in packages, trucks, wagons, containers, etc.;

– appendix IV: pollution by wastewater;

– appendix V: pollution by garbage;

---

37 This convention, one of the oldest in the international Law of the Sea, with versions dating from the early 20th Century (the first from 1914, following the sinking of the *Titanic*), pertains to the safeguarding of human life at sea and contains several provisions that play a role in anti-pollution matters: the double-rudder system that was missing on the *AmocoCadiz*; collision-limiting rules, and the International Safety Management code, or ISM, of November 4, 1993, which deals with both the management of safety aboard ships and the prevention of pollution. If it had been applied aboard the *Erika*, many abstruse management practices would have been prevented.

38 Applicable in any "maritime space under jurisdiction", and thus also in EEZ, this framework convention also contains a whole series of definitions of terms used, including "waste", "ship", etc., as well as some very general rules that are not highly operational, with the most important ones found in the appendices to the convention (see *infra*).

– appendix VI: atmospheric pollution by ships.

The first five appendices date from the same time as the framework convention and the protocols. The first two appendices are compulsory for any State that is a signatory to MARPOL, while the other appendices remain voluntary. Appendix VI was added much later, in 1997. It is optional for States and is part of existing international legislation relative to the fight against climatic change. All of these appendices have entered into force, but some of their amendments still have not.

Each appendix is symmetrically constructed and concerns two angles for the prevention of the pollution to which it is devoted. The first angle concerns the design and fitting-out of ships[39], and the second angle concerns a waste-management system[40]. This system is much more drastic[41] in zones recognized by the IMO, appendix by appendix, as MARPOL special zones[42], with a single objective: making operational pollution as well as the consequences of accidental pollution as minimal as possible[43], or even non-existent.

---

39 Oil tankers must have, if possible, separated ballast, and must be equipped with sloop tanks and continuous waste control systems. They must be double-hulled (or the equivalent) and have a specific size of waste sorter determined by category of waste as well as by standards of labeling, the stowage of dangerous merchandise transported in packages, wagons, or sulfur oxide waste in the air resulting from the use of certain marine fuels, and all of this must be attested to by international certifications, the regularity of which is verified during checks by each port nation.

40 This waste is prohibited on principle except in specific conditions where it is authorized – see *infra*.

41 Contrary, perhaps, to what is often said, the principle of MARPOL's appendix I on hydrocarbons, for example, does not concern the prohibition of waste or deballasting, but rather its regulation (continuous ongoing onboard waste control system that cannot contain more than 15 PPM of hydrocarbons, at a certain distance from the nearest coast only, when the ship is en route, etc.). Conversely, in special zones, a stricter system that may extend as far as the prohibition (for example, in the Antarctic) of all waste is provided for. This regulation of waste makes it compulsory for nations to have land-based facilities in their ports to receive what has not been able to be disposed of, in compliance with international law.

42 These SZs have been proposed for the IMO's decision by a coastal state or states for one and/or other MARPOL appendices; examples of this are the Mediterranean for appendices I and V; the Antarctic for appendices I, II and V; and the North Sea for appendices I, V and VI (SOx).

43 As in the case of the double-hull rule or that of ballasts in defensive locations, for example, as provided for by appendix I in order to minimize the consequences of an accident in terms of pollution, or waste regulations in the context of operational pollution – *infra*.

Ensuring compliance with these technical requirements, as well as with SOLAS provisions and some ILO conventions, is first managed by the flag State, which assumes responsibility in this area and delegates this task to classification companies which then issue international certificates. But, these checks are also, and sometimes especially, carried out by port State controls[44], which have existed in Europe since 1973, and under the aegis of the inter-administrative accord entitled the Memorandum of Understanding (MOU) signed in Paris in 1982. Since its inception, this accord has given rise to numerous emulators worldwide and, since the 1990s, has been compulsory in France under European Union law[45]. A tanker that does not respect one or another of these conventions may be boarded and searched in port, and allowed to leave this port only to travel to a naval shipyard in order to be fitted out in accordance with relevant international standards.

Thus, the MARPOL convention is extremely technical; not only in terms of vessel construction standards, but also with regard to regulations and conditions concerning the management and disposal of any waste that these vessels may introduce into the environment. MARPOL details infractions[46], leaving it to signatory States and coastal and flag States to define the penalties that will be attached to these infractions[47]. However, if the coastal State thus has full jurisdiction to prosecute polluting ships, particularly those

---

44 Also considered in article 218 of UNCLOS.

45 Since European Community directive 95/21/EC of June 15, 1995, subsequently reviewed.

46 The convention actually specifies the conditions of lawfulness of waste and the obligations relative to this waste in terms of the construction, design, and fitting out of ships, chemical tankers, oil tankers, etc. Though oil tankers and chemical tankers can still emit waste in the hypothetical event that this is necessary for the safety of the ship or its passengers or when this waste is part of anti-pollution measures, it is usually prohibited barring provisions to the contrary. For hydrocarbons, for example, appendix I specifies that, outside special zones in which all waste is prohibited except for separated ballast, waste disposal is possible under certain cumulative conditions such as: if the ship is en route, if this waste does not exceed a certain effluent level (15 PPM), an automatic waste disposal stoppage system must be in place, and finally if the ship is at a certain distance from the coast (that is from the baselines), normally at more than 50 nautical miles from the nearest coast. The depth of the water is also taken into account when dealing with chemical tankers as provided for by appendix II, for example. However, the drastic system imposed in SZs has a flip side; it can only be applied in a zone for which it is specified if ports have facilities to receive wastewater from hold waters, cargo hold cleaning waters, etc. If this is not the case, the "general" system is applicable.

47 See in France, provisions relative to this aspect principally contained within the environmental code.

guilty of illegal waste disposal, this jurisdiction can only be exercised if the pollution occurs within 12 miles of territorial waters. Things are also different and more complicated if the pollution occurs in an Exclusive Economic Zone (EEZ). In this case, the flag State has certain rights of legal action exclusive to certain conditions specified by article 228 of UNCLOS which shift pollution repression measures to jurisdictions of nations other than the flag State except in cases of serious pollution. According to article 228, except in cases of serious pollution, legal action must be ceased before the courts of the polluted State, if the flag State has undertaken legal action toward its ship within a certain deadline; if it organizes deterring sanctions in the matter; and if this State is trustworthy in its desire to effectively prosecute its polluting vessels. This article was recently emphasized before the French courts[48] and has often been brought up subsequently, expressing the anger of polluted populations and marine environmental protection organizations, especially when the penalties imposed by the legal system of the flag State are less severe, and thus less dissuasive, than those imposed by French jurisdiction[49].

Other forms of pollution by ships have been the subject of media coverage recently, and have attracted IMO regulatory efforts. For years, it has been planned to add new appendices to the MARPOL convention, following the example of what was done for appendix 6 relative to atmospheric pollution by ships, the only appendix that does not date from the drafting of the initial convention. Yet, in matters of these other sources of marine pollution by ships, the IMO has chosen to enact separate conventions that have no link to MARPOL. Such is the case for the issue of the introduction into marine waters of a foreign or exogenous living organism trapped in the ballast waters of ships making international voyages; for antifouling paints, and for the recurring issue of ship recycling; all

---

48 See, for example, the case of the *Transarctic*, a Norwegian ship that disposed of hydrocarbons in the French EEZ in 2005 [LEM 06].

49 For example, the *Vytautas*, a Lithuanian ship involved in illegal waste disposal in the Atlantic, was punished by its home country by a fine of 22,634 euros, though in this case the Brest Tribunal had handed down a fine of 700,000 euros in the first instance. The appellate court of Rennes was, therefore, obliged on January 20, 2011 to pronounce the discontinuation of legal proceedings before the French courts for this case, causing certain parties to denounce this action as "when Lithuania overwrote the prices". Position of the Court of Appeal on this point in accordance with article 228 of UNCLOS: Court of Appeal, Ch. Crim. May 5, 2009, 2 copies, Bull. crim. 2009, no. 85.

significant sources of pollution which are the subject of IMO conventions that have not yet all gone into effect.

3.2.3.2. *Taking new pollutants into account: waiting for the entry into force of certain pertinent IMO conventions, or the awkward realm of soft law at the international level*

The issue of pollution by ships appeared on the legal horizon in the face of disasters such as black tides. It is no surprise, therefore, that oil and chemical tankers are the primary vessels targeted and regulated by law, due to the dangerous nature of the cargoes they carry. However, MARPOL already considers forms of pollution that have no relationship to cargo: wastewater, garbage, etc. It is necessary to acknowledge that all ships – whatever their cargo – are liable, due to the very fact of being ships, to cause pollution by means other than their cargoes. The regulation of these "other" pollutants has been undertaken by the IMO as well as certain regional international organizations, such as the European Union: invasive species (section 3.2.3.2.1), antifoulings (section 3.2.3.2.2) and the very general but crucial issue of recycling ships at the end of their lifecycles (section 3.2.3.2.3). In these three cases, intended to provide a more complete response than the one given by the SOLAS and MARPOL conventions, for example, as with the IMO's[50] promotion of the *cradle to grave*[51] objective for new ships, the existing conventions have not yet all gone into effect at the international level, which does not prevent certain States or organizations[52] from setting local standards or using guidelines and other non-binding IMO circulars, embedded in this case in soft law, despite repeated calls by the IMO secretary-general to ratify these conventions as rapidly as possible see below.

3.2.3.2.1. Prevention of the introduction of exogenous organisms or invasive species

Article 196 of UNCLOS very clearly states that "States must take all necessary measures to prevent, reduce, and control marine environmental pollution [...] resulting from the introduction of new or foreign species liable to provoke significant harmful changes".

---

50 See the IMO's 2012 contribution to the Rio + 20 Summit and relative to the Concept of a sustainable maritime transportation system.
51 In French, *"du berceau au tombeau"*.
52 See, for example, the provisions of European Union law relative to the recycling of ships, *infra*.

There are several ways of introducing foreign species into a marine environment, not all of which necessarily constitute pollution. This is the case, for example, with the natural displacement of some species due to the opening of canals connecting two maritime ecosystems that had been previously separated by nature, or with repopulation when a species has been eradicated and efforts are made to revitalize the economy by introducing an equivalent foreign species[53]. Navigation is indisputably the most concerning activity of this type for the international community; even though only 3% at most of living exogenous species displaced in this way adapt to their new environment, there are some that completely destroy the local biodiversity and ecosystems into which they are introduced. Vulnerable Australian mariculture zones were especially victimized by cases such as this in the 1990s, which also saw the addressing by various national and international authorities of the issue of "marine pollution" as considered by article 196 of UNCLOS, mentioned above[54]. The IMO had no choice but to submit, particularly under pressure from Australia and Canada, which were especially affected, and in 1991 established the first guidelines[55] on the subject by inciting ships to keep a log of ballast water shifts, changes of ballast water in seas more than 2,000 m deep, etc. These guidelines were not legally binding, however, and on February 13, 2004 the international community established a convention relative to the control and management of ballast water and sediment by ships (called BWM, for Ballast Water Management) which made most of the provisions contained in the 1991 guidelines legally binding but adapted them with regard to possible technical advances in the area. Unfortunately, this BWM convention has not yet been made effective, and the guidelines, which have been modified, remain applicable only at the global level, and are still optional, leaving room for national regulations currently under discussion[56].

---

53 See the case of oysters in France, of Portuguese origin, subsequently decimated and replaced by oysters of Japanese origin in the 1970s.
54 When freshwater species were found in the Saint Lawrence seaway in Canada, Japanese starfish in Australian waters or European green crabs in South Africa; but, these waters also carried viruses such as cholera and micro-organisms harmful to human health such as *Alexandrium* from the Chesapeake Bay, a microalgae toxic to humans that infested mussels in the 1990s in the waters off Côtes-d'Armor in France (see [LET 11]).
55 Specified in November 1993 by General Assembly Resolution A 774 (18) and then in 1997 by Resolution A 868 (20).
56 See, for example, the new American regulations on the subject, called the Vessel General Permit (VGP), effective date December 20, 2013 [LEM 14].

3.2.3.2.2. Prevention of pollution by antifoulings

Antifoulings are applied to the hulls of ships to ensure safety, the maneuverability of the ship, protection against corrosion, etc. This paint formerly contained small amounts of tributyltin (TBT), a substance which, accumulating in the water, rapidly proved biocidal[57]. Considered a polluting substance and prohibited as such, antifoulings containing TBT were not included in existing regulations, notably at the European level. Denouncing this paint in the 1992 Rio Agenda 21 as a significant pollution issue, on October 5, 2001 the IMO introduced an international convention relative to the control of harmful shipboard antifouling systems, called the Anti-Fouling System (AFS) convention. This convention prohibits the use of any harmful organotin compounds in marine paints as well as the use of other harmful substances for antifouling purposes. The AFS convention was expected to go into effect in 2002, and the European Union adopted regulation 783/2003/EC on April 14, 2003 to ensure its initial application at the regional level. Subsequently, save for the application of local or national rules, only the IMO's recommendations on the subject were applicable at the international level until the AFS convention finally took effect on September 17, 2008.

3.2.3.2.3. The question of recycling ships at the end of their lifecycle

It goes without saying that, in the context of analyzing the lifecycles[58] of ships and the "cradle to grave" concept embraced by the IMO, the question of recycling ships at the end of life must come up. Indeed, this has been included as a requirement among IMO provisions (particularly with regard to double-hull oil tankers) due to the fact that most maritime accidents have been due in the past to the advanced age of the ships involved. All of this has also resulted in the refreshment of some fleets, necessitating the recycling of old ships that no longer meet standards. End-of-life ships can be legally treated as garbage, and the 1989 Basel convention on cross-border transport of waste could be partially applied, but there were no rules specific to the recycling of ships that could not end their lives in any other way (as breakwaters, for example). The IMO, in its directives[59] relative to

---

57 See in particular the studies conducted on shellfish in the Arcachon basin in France, for example.

58 For more information on lifecycle analysis, see [VOI 14].

59 Adopted by Resolution A 962 (23) and modified in particular by the 2005 IMO resolution A 980 (24).

recycling, modified in 2005, reiterates that, while the principle of recycling ships is a good one in itself, the labor practices and environmental standards observed in recycling facilities often leave much to be desired[60]. However, these directives are not legally binding and should be replaced by a compulsory text.

The question of ship recycling, then, was one of the fundamental elements of sustainable development when the IMO made its 2012 contribution, during the Rio + 20 international summit, relative to the concept of a sustainable maritime transportation system. This legal void concerning recycling was filled by the Member States of the IMO with the adoption on May 15, 2009 of the Hong Kong convention for the safe and environmentally sound recycling of ships. With a view to encouraging the entry into force of this convention and emphasizing the importance of it, and in view of the fact that only three nations, including France on July 2, 2014, have ratified it to date, the European Union has recently put forth a regulation relative to this issue[61], which includes a list of substances and materials prohibited aboard new vessels and supplying a European list of recycling facilities worldwide corresponding to the environmental criteria of the Hong Kong convention.

Despite all these provisions implemented mainly in order to prevent operational pollution, it is undeniable that all these measures, even Marine Traffic Organization (MTO), cannot avoid accidents and pollution in every case, though MARPOL regulations are intended to reduce the consequences of accidental pollution. This leads to the question of intervention on a ship posing a threat to the marine environment.

3.3. Intervention in the event of accidents or risk of accidents

Accidents at sea occur frequently and may, whether the ship is substandard or not, lead to more or less major pollution, and notably to black tides, if the ship's cargo is composed of hydrocarbons.

---

60 See on this subject the tribulations of the French ship *Clémenceau* [LET 09].

61 (EU) regulation no. 1257/2013 of the European Parliament and Council of November 20, 2013 relative to the recycling of ships and modifying (EC) regulation no. 1013/2006 and directive 2009/16/CE, JOUE no. L 330 of 12/10/2013, p. 0001-0020.

Thus, the IMO, after first envisioning the ability of a third-party State to intervene on the high seas in the place and without the prior approval of the flag State (section 3.3.2), eventually made it compulsory for shipowners, States, port authorities, etc., to provide emergency mechanisms to be used in the event of accidental pollution (section 3.3.1).

3.3.1. *Preparedness via the OPRC convention*

Discussing preparedness here is somewhat remarkable, considering that the *Oil Pollution Preparedness Response and Cooperation* (OPRC) only dates from November 30, 1990, though the international desire to cooperate in the protection of the oceans goes back to the 1920s. In actuality, the international community did not begin by establishing rules requiring anti-pollution equipment in case of incidents on board oil or chemical tankers, as the SOLAS convention did when defining obligations relative to lifesaving equipment in 1914. These rules would not be set for oil tankers until 1990 by the OPRC convention, which was supplemented in 2000 by a protocol relative to noxious and potentially dangerous substances (the HNS protocol), which went into effect on June 14, 2007.

The 1990 convention required signatory States to create regional and national emergency control plans (also in compliance with the recommendation made by article 199 of UNCLOS) to be put into action in the event of incidents causing marine pollution. Such plans already existed in some States[62] and maritime regions, notably as part of conventions on the protection of "regional seas" established under the aegis of the United Nations Environment Program (UNEP)[63]. However, the OPRC convention also requires emergency plans on board ships and in the ports of signatory States – a very new development.

This convention also fills in an international legal void that had previously been filled only by the acknowledged ability of coastal States to

---

62 See the ORSEC MER (POLMAR) plan in France, reviewed by the civil safety modernization law of August 13, 2004.

63 As in the case, for example, of the "UNEP conventions on regional seas", all of which contain an article concerning information and cooperation in the event of a critical situation in the maritime zone concerned and are supplemented by a protocol for counter-measures in the event of a critical situation, such as in the Mediterranean, called since its 2002 version the "Prevention and critical situations" protocol, as well as for east Africa and the regions of the Caribbean and the Persian Gulf.

intervene aboard foreign vessels threatening their coasts (this allowance is of course still in existence). The OPRC convention completes the provision established by international law, though the order of these steps was somewhat irregular.

3.3.2. *From the 1969 IMO convention on intervention to article 221 of UNCLOS*

The 1967 *Torrey Canyon* disaster was indisputably responsible for the increased awareness of the necessity of developing an international legal corpus beyond OILPOL, the existing prevention convention at the time. One of the first elements of this corpus was the 1969 IMO convention on high-seas intervention[64]. The oil tanker Torrey Canyon sank off the southern coast of Great Britain in the open sea north of France; that is in a space where, legally, only the flag State – Liberia in this case – was allowed to intervene on board the ship on the basis of the applicable law of the time. Because the oil tanker's cargo, leaking into the sea, posed a threat to the French and British coasts, authorities in these two States (though they were not competent to do so under the Law of the Sea) made the decision to intervene on the high seas aboard the foreign vessel, which they chose to sink in order to avoid serious damage to their respective coasts and to related activities by their own nationals. It was decided to retroactively approve this infringement upon the Law of the Sea via an IMO convention which, by its universal nature, was the only body authorized to modify the principle of freedom of the seas on this point, and consequently impose the non-interference of the flag State in cases of imminent danger of hydrocarbon pollution caused by an accident and threatening the shores of one or more coastal States and their relevant activities. This convention was often used without ever requiring the coastal State to intervene, as well as without allowing the flag State to object to this interference, considered a case of self-defense, or "self-protection" under international law. In 1978, the disaster involving the *Amoco Cadiz* – though not considered on the basis of IMO texts – caused the emergence of a problem in the course of this intervention, at the same time as the United Nations was debating the content of the future UNCLOS. France then proposed the addition to part XII of this convention on the Law of the Sea of

---

64 The exact title of which is the "International convention on intervention on the high seas in the event of an accident causing or liable to cause pollution by hydrocarbons", IMO, Brussels, November 29, 1969.

an article it had written itself: article 221, which is applicable today within signatory States of UNCLOS, which reuses most of the 1969 convention on intervention while refining it in terms of efficiency.

According to the currently applicable article 221:

> No provision included in this part will infringe upon the right possessed by States under both traditional and conventional international law to take and cause to be applied beyond territorial waters measures proportional to the damage they have sustained, or by which they are threatened, in order to protect their coastline or related interests, including fishing, against pollution or the threat of pollution resulting from an accident at sea, or from acts linked to an accident at sea, which may reasonably be expected to have harmful consequences.

> For the purposes of this article, "accident at sea" should be taken to mean a collision, sinking, or other navigation incident or event occurring on board or outside a ship causing material damage or an imminent threat of material damage for a ship or its cargo.

This article thus enables a coastal State off the coast of which, beyond the 12 miles of territorial waters, including in an exclusive economic zone, a polluting accident occurs, to intervene in order to limit the consequences of the situation on its coasts. It also – and this is the innovation introduced by the French version in comparison to the IMO convention on deep-sea intervention and to traditional maritime practices respecting the competence of the flag State – to "cause commensurate measures to be taken". The intention of this new addition was to make conventional, and thus legal internationally, a contractual practice that certain tugboats, particularly French ones, implemented in the case of an accident involving assistance provided to a tanker: the Lloyd's Open Form (LOF) of 1983, and then of 1990. Its goal was to allow a threatened coastal nation to do what it could not truly do legally up to that point: impose, rather than simply proposing, assistance measures it judged necessary, such as the obligation to accept forced assistance, which theoretically, in the case of the Amoco Cadiz, would have prevented the black tide that followed due to overly long negotiations on the amount of the payment due for assistance between the ship and the tugboat. Long criticized and not truly addressed by the 1989 London convention on assistance, this possibility was adopted by a number of States, including its staunchest opponent, the United Kingdom, following the Erika disaster in 1999.

In spite of all these regulations, even when refined to enable maximum efficiency against both pollution and the aggravation caused by this pollution, it often occurs that damages are inflicted for which legal reparations must be made.

3.4. Reparations in the event of damage caused by pollution

France, with its Atlantic coast in particular, is well placed to be aware that accidents happen and that black tides may reach its coasts, often at night, despite all the precautions that may have been taken to prevent a catastrophe from occurring.

There are occasions, therefore, when the question of reparation for damages caused arises. The responsibility incurred may be criminal, arising partially or wholly from domestic[65] or European law in matters involving European Union Member States. International law has developed mainly instruments of civil responsibility. Though this existing international law is not limited to the reparation of damages by hydrocarbons, only conventions having to do specifically with this are currently in force, and thus applicable (section 3.4.1). These are the conventions, still extant in their 1992 version, on the civil responsibility of the owners of ships transporting hydrocarbons (CLC 1992, or Civil Liability Convention of 1992) and on the International Oil Pollution Compensation Fund (IOPC 1992), with the former preceded by a 1969 CLC convention and the latter by a 1971 IOPC convention which they are aimed at improving, and through which the IMO applies the "polluter pays" principle. Some additional modifications have recently proved necessary[66] (section 3.4.2).

3.4.1. *The prioritizing of reparations for pollution by hydrocarbons*

This is undoubtedly the vestige of a system of law developed in reaction to an event, but it is indisputable that international law on marine pollution began its focus on reparations for marine pollution by hydrocarbons after the symbolic 1967 disaster involving the *Torrey Canyon* and the other frequent

---

65 See the landmark case of the *Erika* disaster, where criminal proceedings took place, but with the constitution of civil parties.
66 In 2001, the IMO also developed the convention on civil responsibility for pollution-related damages caused by hold hydrocarbons. This convention went into effect on November 21, 2008; it will not be discussed here.

and highly visible black tides that followed. The cleanliness of oil is self-evident; there are black globules on beaches, and birds stuck in the open sea, and both authorities and the public have fully processed these sights[67]. The issue of pollution by hydrocarbons has, therefore, been prioritized ("all that black on all that white", sighed Professor Martine Rémond-Gouilloud [REM 89] after the Exxon Valdez disaster in Alaska), including matters of reparation. While other substances, mostly chemical but not entirely (palm oil, etc.), are also dangerous for both ecosystems and human health, they are not – as widespread as they are, and often even more dangerous – as "spectacular".

There is another convention that addresses reparation for noxious and potentially dangerous substances (the HNS convention, for Hazardous and Noxious Substances) other than the hydrocarbons targeted by the OMI CLC, modeled after conventions having to do with hydrocarbons. However, the HNS convention has not yet gone into force, though preparatory work has continued since the 1960s, and though it was signed in 1996 and a protocol negotiated in 2010 is intended to help make this entry into force happen as soon as possible due to modifications of the initial text. It is now a matter of ratifying the HNS 2010 convention. It is self-evident that the international community is regularly bothered by concerns relative to the risks posed by this "invisible tide", but there are technical questions posed that trouble certain States whose consent is needed to ensure the entry into force of the HNS convention. France considered ratifying the convention at the time of the October 31, 2000 sinking of the Italian chemical tanker *Ievoli Sun* off the coast of The Hague, the cargo of which included 4,000 tons of styrene[68]. The question of a possible catastrophe led the French government to consider ratifying the convention, which it did not do in the end because the threatened marine environmental disaster involving the Italian ship did not occur[69]. At a time when environmental principles of prevention and

---

67 This perspective may be used to approach the wide-ranging interest in green-algae pollution, due more to its visibility than to its danger (which is not being called into question) for the marine environment.

68 Chemical compound used in manufacturing plastics, notably polystyrene (Styrofoam). This chemical substance is not considered very dangerous for the environment, since it does not bioaccumulate greatly or persist to a great extent in the natural environment. Though it is toxic, styrene is not listed among the most toxic products by the IMO (product category Y according to MARPOL's nomenclature).

69 In reality, after the ship's sinking, virtually all of the products were able to be pumped without spilling into the environment. The low danger of the main product present on board, combined with the small quantities released into the sea, prevented serious chemical pollution in the English Channel.

precaution are being validated and promoted, States are clearly still having difficulty being prepared for disasters; these States include even France, which spearheaded the work to modify the HNS convention in 2010 and has ratified it since then, but without the effect of bringing it into force.

For these reasons, only conventions relative to reparation for damages in the event of pollution by hydrocarbons are currently possible; the black tides that follow these polluting events (and the lack of major chemical polluting events) have rendered it so.

3.4.2. *The IMO Civil Liability Convention and FIPOL 1992*

With regard to responsibility and reparation, we must still identify the damages that are considered eligible for reparation by mobilizable international conventions (section 3.4.2.1) before examining how this responsibility is framed and limited (section 3.4.2.2).

3.4.2.1. *Reparable damages*

The position of the law with regard to responsibility and reparation for damages is that reparations are made only for direct, assured and assessable damages.

Reparation for "anthropocentric" damages, or damages caused to humans[70], their goods[71], or their activities[72], has never been truly problematic in comparison to reparation for patrimonial damages[73], because anthropocentric damages can be assessed more or less easily, and their certainty attested to[74].

---

70 Intoxication, for example.
71 A ship stuck in oil in a port, for example.
72 Fishing, tourism and hotels, etc. affected for example.
73 Meaning able to be "billed for", even if they are moral.
74 The question of their directness has been problematic at times. FIPOL refuses payment for so-called "second-degree" damage, as it did in the case of the *Erika* disaster, for example, which involved the request by the owners of a commercial site at Belle-Ile at sea, for the cancelation by the lessee of the rent for this site for the year 2000 (the Lebaupain affair), as in other cases of patrimonial damage caused to shellfish merchants geographically located too far from the site of the catastrophe (the *Sea Empress* disaster off the coast of Wales in 1996, for example).

Thus, anthropocentric damages are naturally the only ones currently indisputably recognized by IMO conventions on the subject as compensable, under the conditions provided for by these conventions[75].

This is not the case for ecological damage; that is damage caused to the environment as such, which has often been considered uncertain and non-assessable, and thus non-compensable except sometimes symbolically[76] or by application of mathematical equations and other fixed solutions, but these are principally at the national level[77].

Moreover, this ecological damage is not precisely addressed by international texts relevant to the subject.

According to article 1.6 of the 1992 CLC, "Damage by pollution means:

a) harm or damage caused outside of a ship via contamination arising as the result of a leak or expulsion of hydrocarbons from the ship, or which this leak or expulsion produces, it being understood that reparations paid for environmental changes other than lack of earnings due to these changes will be limited to the cost of the measures reasonably required for restoration that have been or will be implemented;

b) the cost of safeguarding measures and other harm or damage caused by these measures".

Article 1.7 specifies that: "Conservation measures refer to all reasonable measures taken by any person after the occurrence of an incident in order to prevent or limit pollution".

It is expressly stated that environmental damages are not considered to fall under the definition of "damage by pollution" provided by the CLC and

---

75 See the compensation manual developed by FIPOL on this point.

76 See its reparation of a symbolic franc in the so-called "red mud" affair following the sinking off the Corsican coast of substances by the Italian company Montedison, TGI Bastia 8 December 1976, *Prud'homie des pêcheurs*, Dalloz, 1977, 427, note M. Remond-Gouilloud.

77 Sometimes by application of fixed formulas, as in the case of an event polluting the Baltic Sea in 1979, in which the solution of the Tribunal of Riga regarding reparations for ecological damage set at one ruble per m^3 of polluted seawater, while the criminal court of Toulon refused in the 1980s to pay for ecological damage caused by illegal sea urchin fishing in the national park of Port Cros on the basis of market prices, with the judge declaring that "sea urchins provide a more important ecological service in the event" than the one paid for on this type of economic basis.

also used by the 1992 FIPOL convention, except with regard to reasonable restoration and patrimonial costs[78]. This solution, according to FIPOL officers, does not have adequate financial resources to cover these damages via State contributions except where a political decision to the contrary is made. If States wish to make reparations for ecological damage, they must only change the definition of damage by pollution[79] given by the texts and accept an increase in their contribution to FIPOL, which only functions as an insurance "mutual fund", or something very similar to it.

However, this issue continues to receive extensive media coverage, and numerous cases of national legislation and jurisdiction are making allowances for the acknowledgement and assessment of and compensation for environmental damage in addition to CLC/FIPOL reparations.

This type of damage raises specific questions in terms of responsibility with regard to its certainty, its assessable nature and above all the identity of its victim, which is not a person *a priori*, but rather an animal, vegetable, or mineral, etc. Responsibility is perceived above all in law as a relationship between two individuals (a victim and a culprit, or a creditor and a debtor). Yet, social demand has compelled the application of the classic model of responsibility to ecological damage, for better or worse. States including Italy and Russia have issued legislation since the late 1970s authorizing reparations for ecological damage, and some national jurisdictions have done much the same; the very recent and unprecedented position of the French Court of Cassation[80] on the *Erika* disaster is a particularly striking example of this.

---

78 This "reasonable" character has been the subject of an assessment by FIPOL operating under the theory that cleaning and restoration efforts sometimes create more damage than they repair (the "How Clean is Clean?" theory). It is sometimes asserted, as part of the still highly esteemed theory of assimilative capacity regarding the marine environment, that there is a reason to let nature take its course and "do its own work powerfully," as decreed by the judgment in the case of a black tide in a mangrove swamp in Puerto Rico, United States, in 1973, caused by the oil tanker *Zoe Colocotroni* (D. 1982 Chron. 33 M. Rémond-Gouilloud), and that cleanup efforts may create more damage than self-reconstitution, and that they may thus constitute an "unreasonable" conservation measure, which is, therefore, not compensable.
79 Or even gamble on an evolution of "jurisprudence" with regard to FIPOL: see the Grenelle of the Sea, "Mission FIPOL" 2010, a theory not refuted by the secretary-general of this fund.
80 Cour de Cass. Ch. Crim. 25 September 2012 no. 3439 on the *Erika*, in which certain parties were recognized as compensable victims of this environmental damage, including some environmental protection organizations affected in their *animum societatis* – see [KEL 08] – and affected coastal regional authorities (departments and municipalities); for the decision of the court of appeal, see [DEB 13].

3.4.2.2. *Closely supervised and still-limited reparations*

The objective of developing the Civil Liability Convention and the FIPOL convention is to promote and facilitate reparations for damages caused for victims, but without guaranteeing total reparations in every case. This is an "old" custom of commercial maritime law according to which, because maritime expeditions include a high degree of risk by their very nature, the responsibility of operators is limited in terms of the amount of reparations they must pay for damages caused[81] if these damages existed during the expedition. The same is true for the transport of dangerous merchandise such as hydrocarbons, and more broadly for marine pollution caused by ships.

This system of reparations is organized into two stages:

– First, reparations are due to be paid by the owner of the ship responsible for the pollution, who is often insured, since insurance is compulsory for ships of more than 700 grt. Responsibility falls objectively on the "owner" of the ship regardless of the actual fault of this owner[82]. Conversely, the CLC specifies that the owner's responsibility to pay cannot, whatever the assessment of the amount of damage caused, exceed a certain limit[83], based on what is called "limitation of responsibility" in maritime law.

– Next, if the assessed damage exceeds the CLC limit, responsibility falls upon the loaders (oil companies that have oil transported by ship) to pay an additional contribution to these reparations. This additional amount, paid by FIPOL, an international organization in its own right, financed by contributions by signatory States[84], is limited in its turn[85] and has proven insufficient to cover the larger and larger amounts of damage caused. It was

---

81 See, for example, the convention on limitation of responsibility for owners of ships of October 10, 1957 – a theoretically controversial system of limitation in modern times; see *infra*.

82 It will fall to the latter, if necessary, to take recourse action against any possible responsible parties to blame for the pollution.

83 The CLC limit is 4.51 million SDR, or approximately 5 million euros. This limit does not apply in the case of inexcusable transgression by the owner; see *infra* on this point.

84 States receiving a certain quantity of hydrocarbons by sea and which often, via taxation (for example, TIPP) put together this contribution from large quantities of oil imported by ship by oil companies into their territory.

85 The FIPOL 1992 limit and the CLC limit equal 203 million SDR, or 228 million euros. Only 85% of some 7,000 requests for compensation for damages caused by the *Erika* have been paid under these terms, for example, with France prioritizing individuals and regional authorities in the matter of reparations.

decided, immediately after the *Prestige* disaster off the Galician coast in November 2002, to develop a third convention that would enable full reparation for damages caused by a black tide. This was accomplished on May 16, 2003 with the creation of a fund called the "Supplementary Fund" or FIPOL II, which is attached to FIPOL 1992 as far as its signatory States are concerned but remains legally separate.

Thus, in the event of compensable damages caused by a black tide, compensation can occur at three levels. The owner – or, more precisely, his/her insurer, if applicable – is the first to pay. If the overall amount of the damages exceeds the maximum amount of reparations set by the CLC and paid by the owner, victims can obtain an additional amount from FIPOL[86], the amount of which is also limited. For signatory States to the supplementary fund that became effective on March 3, 2005, and if the FIPOL 1992 supplement is still not enough to cover full reparations, victims are able to mobilize this third level of compensation[87]. This maximum compensation can reach the equivalent in SDR[88] of 900 million euros[89].

It is clear, then, as we conclude this presentation of the international system of measures against marine pollution by ships, that the IMO plays a crucial role in ensuring uniform protection of the environment, which agrees with its stated goal in 2012 to develop "the concept of a sustainable maritime transport system".

---

86 Which can also be directly and solely solicited if the shipowner cannot be found or is insolvent.

87 It should be noted that while limitation of responsibility for maritime operators is a traditional provision in maritime law, its continued existence is highly controversial, both in matters touching pollution and in other maritime claims. In fact, doctrine increasingly advocates the elimination of this concept of limitation, and of the difficulty of making it non-applicable to an incident. Thus, with regard to the CLC, this limitation cannot be actionable in the case of inexcusable fault (difficult to prove), and it is often suggested that this exception be changed due to the fact that simple fault would make the limitation non-applicable and would then require full reparation for damages. See, for example, the "Mission FIPOL" proposal from the 2010 Grenelle of the Sea (p. 8) relying on the jurisprudence of the European Court of Justice, also used by the Court of Appeal of the municipality of Mesquer in its decree of September 17, 2008.

88 For "special drawing rights," which is the international currency used by the International Monetary Fund (IMF) defined on the basis of several currencies.

89 Which approaches the full assessed amount of the damages caused by the black tide from the *Prestige*.

Yet, it is evident that, despite well-executed achievements significant efforts in terms of application, often including ratification to enable the entry into force of international legal instruments, remain to be made.

In the Biennum, its strategic plan for 2014–2019, the IMO reiterates that "the mission of the International Maritime Organization (IMO), as a United Nations specialized agency, is to promote safe, secure, environmentally sound, efficient and sustainable shipping through cooperation. This will be accomplished by adopting the highest standards of maritime safety and security, efficiency of navigation and prevention and control of pollution from ships, as well as through consideration in the related legal matters and effective implementation of IMO's instruments, with the view of their universal and uniform application[90]".

It remains the case that though tireless efforts must continue to ensure the effectiveness of the international environmental provisions of the IMO, the organization must also carry out this mission in a context of particularly harried globalization, in which the fundamental human question of safety (piracy, terrorism, etc.) competes for attention with maritime security and, due to its urgency, may supplant it at times, further delaying progress in the protection of the sea from ships in favor of more effectively protecting humans at sea from the actions of other humans.

3.5. Bibliography

[BOI 98] BOISSON P., *Politiques et droit de la sécurité maritime*, Bureau Veritas, Paris, 1998.

[DEB 13] DEBORDE A., "L'apparition de la notion de préjudice écologique en droit français", *Revue Droit de l'Environnement*, no. 18, 23 July 2013.

[KEL 08] KELIDJIAN F.X., "L'indemnisation du préjudice écologique à la suite du jugement Erika: le cas de la LPO", *DMF*, no. 695, pp. 780–793, 2008.

[LEM 06] LE MARIN, Convention de Montego Bay: Vers une fuite judiciaire des pollueurs, p. 8, 13 January 2006.

[LEM 14] LE MARIN, Eaux de ballast: la réglementantion américaine est entrée en vigueur, p.5, 5 February 2014.

---

90 IMO strategic plan for 2014–2019, Res. A/1060 (28) of January 27, 2014.

[LET 09] LE TÉLÉGRAMME DE BREST, L'ex-Clem en dix chapitres, p. 15, 26 January 2009.

[LET 11] LE TÉLÉGRAMME DE BREST, Journées conchylicoles IFREMER – Résumé, p. 13, 23 June 2011.

[MON 14] MONACO A., PROUZET P., *Value and Economy of Marine Resources*, ISTE, London, and John Wiley & Sons, New York, 2014.

[REM 77] RÉMOND-GOUILLOUD M., *Prud'homie des pêcheurs – Affaire du Montedison*, 427, note, Dalloz, Paris, 1977.

[REM 82] RÉMOND-GOUILLOUD M., *Affaire du Zoé Colocotroni*, Chron. 33, Dalloz, Paris, 1982.

[REM 89] RÉMOND-GOUILLOUD M., *Du droit de détruire – Essai sur le droit de l'environnement*, PUF, Paris, 1989.

[REM 91] RÉMOND-GOUILLOUD M., "Marées noires: les Etats-Unis à l'assaut (l'Oil Pollution Act 1990)", *DMF*, no. 506, pp. 339–353, 1991.

[REM 93] RÉMOND-GOUILLOUD M., *Droit maritime*, Pedone, Paris, 1993.

[VOI 14] VOISIN S., FREON P., "Fisheries and aquaculture sustainbility", in MONACO A., PROUZET P. (eds), *Value and Economy of Marine Resources*, ISTE, London and John Wiley & Sons, New York, 2014.

4

Management and Sustainable Exploitation of Marine Living Resources

4.1. European policy on the sustainable exploitation of marine living resources[1]

Originally, the European Union (EU) – which was known as the European Community until the entry into force of the Treaty of Lisbon on December 1, 2009 – took an interest in the exploitation of marine living resources through the fishing activities of its member states. As we will see, the sustainable character of fishing did not truly become a concern for the EU until the 1990s, with the introduction of requirements for environmental conservation[2] (see section 4.1.1). This evolution of European policy on the exploitation of marine living resources is in line with respect for the major principles (see section 4.1.2) and decision-making mechanisms that exist within the EU, which are particularly delicate to implement, in matters pertaining to the exploitation of marine biological resources (see section 4.1.3).

Chapter written by Annie CUDENNEC and Olivier CURTIL.
1 This study will involve only marine fishery products and will not address products from the aquaculture industry.
2 The EU is a member of the United Nations Convention on the Law of the Sea (UNCLOS), signed at Montego Bay on December 12, 1982 and made effective on November 16, 1994. This convention, which is sometimes considered as the true "Constitution for the oceans", establishes a general framework for the management of the oceans, and the EU's actions are in accordance with the prescriptions of this convention in matters pertaining to the conservation of marine biological resources.

4.1.1. *The European Union and the sustainable exploitation of marine living resources: a long and complicated history*

4.1.1.1. *A brief history*

It was not until 1966, 9 years after the signing of the Treaty of Rome creating the European Economic Community, that the European Commission took an interest in the fishing activities of its member states[3].

At that time, fishing products were still considered as agricultural products[4], and European fishing policy was part of the Common Agricultural Policy (CAP). The Commission did acknowledge, however, that "fishing constitutes a distinctive product of states' agricultural sectors", in that fluctuations in production are wider than those recorded for most agricultural products, and the investments required for the exploitation of marine resources are often far and away greater than those needed for small- and medium-sized agricultural businesses. Moreover, most marine species do not have a constant habitat; rather their habitats fluctuate according to the nature of the sea floor, meaning that production depends on "permanent and costly" search efforts. Finally, fishing activity retains an "originality" that is proper to it, which translates in human terms into a "particularism common to all seafarers, born out of the uncommon life they share and the perpetual risks to which they expose themselves at sea".

From then on the Commission has advocated the development of a common policy specific to fishing and separate from common agricultural policy, though still based on the same legal principles. This policy, grounded in an economic perspective, would be aimed at: "increasing the productivity of the fishing industry by developing technological advances, ensuring rational development of production and the optimum use of production factors, ensuring an equitable quality of life for the maritime population..., stabilize markets, guarantee the safety of supplies, and ensure reasonable prices in deliveries to consumers[5]".

---

3 See the report "on the situation of the fishing sector in the member states of the EEC and the basic principles for a common policy", developed by the European Commission and used in part by the Economic and Social Commission, OJEC no. C 58 of March 29, 1967, p. 861.
4 On this point, see article 38 of the TFEU (Treaty on the Functioning of the European Union).
5 Report "on the situation of the fishing sector in the member states of the EEC and basic principles for a common policy", cited above, OJEC no. C 58 of March 29, 1967, p. 863.

The conservation of resources is part of a productivist type of logic: "the improvement of yields and the rationalization of production involves the protection of stocks of natural resources in order to ensure conservation, renewal, and growth[6]".

In line with this, the first regulations relative to a common fisheries policy (CFP), adopted in 1970, were aimed above all at organizing the common market for fishing products and at guaranteeing equality of conditions of access and exploitation of seafloors located in waters falling under the sovereignty or jurisdiction of member states[7].

It would be 13 more years, however, for the first measures on conservation and management of marine biological resources to be adopted, on January 25, 1983 thus creating "Blue Europe", as was frequently asserted at the time[8]. Subsequently, various reforms of the CFP gradually incorporated requirements for environmental conservation in order to make the exploitation of marine biological resources more sustainable, and the current basic regulation, (CE) 1380/2013[9], which went into force on January 1, 2014, reaffirmed that "the CFP guarantees that fishery and aquaculture activities will be environmentally sustainable in the long term" (article 2-1).

It is clear that European policy on the exploitation of marine biological resources has evolved extensively through the years; however, it must still and always respect the major principles of EU law.

4.1.1.2. *A policy embedded in the major principles of European Union law*

The EU is based on founding principles set forth in treaties, within the context of which all European policies are based. The CFP is not exempt from this. Our objective here is not to provide an exhaustive review of these

---

6 See the Commission report, cited above, OJEC no. C 58 of March 29, 1967, p. 865.

7 Council regulation (EEC) no. 2341/70 of October 20, 1970, relative to the establishment of a common policy of structures in the fisheries sector, OJEC no. L 236 of October 27, 1970, p. 1 and Council regulation (EEC) no. 2142/70 of October 20, 1970, relative to the common organization or markets in the fishery products sector, OJEC no. L 236 of October 27, 1970, p. 5.

8 See in particular Council regulation (CEE) no. 170/83 of January 25, 1983 establishing a Community system for the conservation and management of fishery resources OJEC no. L 24 of January 27, 1983, p. 1.

9 European Parliament and Council regulation (EU) no. 1380/2013 of December 11, 2013 on the Common Fisheries Policy, amending Council Regulations (EC) No 1954/2003 and (EC) No 1224/2009 and repealing Council Regulations (EC) No 2371/2002 and (EC) No 639/2004 and Council Decision 2004/585/EC, OJEU no. L 354 of December 28, 2013, p. 22.

principles. We will confine ourselves, rather, to discussing two among them which orient the actions of the EU in a fundamental manner: the principle of conferral of powers (see section 4.1.2.1) and the principle of environmental consistency (see section 4.1.2.2).

4.1.1.2.1. Compliance with the principle of conferral of powers

In accordance with the principle of conferral of powers, which "regulates the delineation of the competences of the Union[10]" (article 5 TEU, Treaty on the European Union), the EU cannot be assimilated into a state, but rather acts, like any international organization, only within the limits of the competences conferred upon it by treaties, in accordance with the wishes of its member states.

It is, therefore, treaties – currently, the TEU and the Treaty on the Functioning of the European Union (TFEU) – that decree the distribution of competences between the Union and its member states. This distribution is simple in appearance, but is in reality much more complex than it seems, since the EU distinguishes various types of competences according to the area of action involved.

First, in certain areas, the EU has exclusive competence. In this case, it alone can "legislate and adopt legally binding acts, which the member states cannot do by themselves unless they are authorized by the Union, or in order to implement the acts of the Union" (article 2-1 TFEU).

The EU has exclusive competence to ensure "the conservation of marine biological resources as part of the common fisheries policy" (article 3-1-c) TFEU)[11]. Thus, the vast majority of measures for the conservation of marine biological resources implemented within the member states of the EU are European regulations.

---

10 Article 5 TEU: (1) The principle of conferral of powers regulates the delineation of the competences of the Union... (2) By virtue of the principle of conferral of powers, the Union acts only within the limits of the competences conferred upon it by treaties in order to achieve the objectives established by these treaties. Any competence not conferred upon the Union in the treaties belongs to the member states.
11 Other areas of exclusive competence of the EU are: the customs union; the establishment of rules of competition necessary for the functioning of the domestic market; monetary policy for those member states whose currency is the euro; and common commercial policy. Finally, under certain conditions, the EU has exclusive competence to conclude certain international accords.

Second, in other areas, the EU has shared competence with its member states. In this scenario, the Union and member states can legislate and adopt legally binding acts. However, the TFEU strictly frames national action by specifying that member states may exercise their competence only when the Union has not exercised, or has decided to stop exercising, its own (article 2-2 TFEU). The CFP, with the exception of marine biology conservation measures adopted within its framework, like those on environmental conservation, result from competences shared by the Union with its member states[12].

In order to manage the complex implementation of competences shared between the EU and its member states, the TEU relies on the principle of subsidiarity. By virtue of this principle, "in areas that do not fall within its exclusive competence, the EU shall intervene only if, and insofar as, the objectives of the action projected cannot be achieved in an adequate manner by the member states, at the central level or at the local and regional levels, but can be better achieved, due to the scope or effects of the projected action, at the Union level" (article 5-3 TEU).

The Union's actions are thus strictly circumscribed: it can act only if its actions prove truly more effective than national action, in accordance with the so-called "test of comparative efficacy".

The treaty clearly defines a subtle interconnection of competence between the Union and its member states that does little to increase the clarity of their respective actions, particularly with regard to the CFP. In reality, it is no easy matter to understand where actions for the "preservation of marine biological resources" as an area of exclusive EU competence cease. In the end, after all, are not all actions taken as part of the CFP intended to conserve resources?

In addition to the principle of distribution of competence between member states and the EU, the implementation of the European policy on the sustainable exploitation of marine living resources requires compliance with another major European principle: environmental consistency.

---

12 Other domains of shared competence, enumerated in article 4 of the TFEU, include domestic markets, agriculture, transport and energy.

4.1.1.2.2. Compliance with the principle of environmental consistency

Remember that in the earliest days of the CFP, environmental concerns were not at the forefront. In 1966, the Commission defined the objectives of the CFP without making any reference to the preservation of the environment: "The goal of the common policy in the fisheries sector is to procure for the population concerned an equitable quality of life by ensuring the individual income of workers through the harmonious and balanced development of the fisheries economy within general economic activity[13]".

It is true that in 1983, the first European regulation establishing a community-wide system for the management and conservation of fisheries resources specified that the goal of this system was "to ensure the protection of fishing grounds, the conservation of marine biological resources, and their balanced exploitation on sustainable bases[14]".

However, it was not truly until 1992, during the United Nations Conference on the Environment and Development (UNCED – Rio) that the community scheme for the conservation of marine biological resources was put into an environmental perspective, taking into consideration the impact of measures taken on the marine ecosystem[15].

Yet, it was only in 2002, with the new regulations concerning the CFP[16], that this policy was finally explicitly bound to comply with environmental

---

13 Report "on the situation of the fisheries sector in the member states of the EEC and basic principles for a common policy," cited above, p. 864.
14 Article 1 of (EEC) Council regulation no. 170/83 of January 25, 1983 establishing a Community system for the conservation and management of fishery resources, cited above. Note that at the time, the European Community did not have express competence to take action in the field of environment, even though an initial plan of action setting up the framework of the community's environmental policy had been established in 1973. However, it was the Single European Act, entered into force in 1987, which included a section expressly dedicated to the environment in the EEC treaty.
15 Article 2 of (EEC) Council regulation no. 3760/92 of December 20, 1992, establishing a Community system for fisheries and aquaculture, cited above. See also the 2nd recital of this regulation: "Whereas the objective should be to provide for rational and responsible exploitation of living aquatic resources and of aquaculture, while recognizing the interest of the fisheries sector in its long-term development and its economic and social conditions and the interest of consumers taking into account the biological constraints with due respect for the marine eco-system".
16 Council Regulation (EC) No. 2371/2002 of 20 December 2002, on the conservation and sustainable exploitation of fisheries resources under the Common Fisheries Policy, OJEC no. L 358 of December 31, 2002, p. 59.

conservation requirements and to apply the principle of environmental consistency, which was subsequently included in treaties, by virtue of which "requirements for the protection of the environment must be included in the definition and implementation of the policies and actions of the Union, particularly in order to promote sustainable development[17]".

In this context, the basic regulations of the CFP form the foundations for the policy of resource conservation and management based on precautionary and ecosystemic approaches[18], both taken from environmental logic.

Today, any measure adopted concerning the management and exploitation of marine biology resources is required to fall in line with this perspective of environmental sustainability, "a precondition for the achievement of overall sustainability[19]". The CFP's contribution to the preservation of the marine environment in its entirety is also clearly confirmed in the "marine strategy framework directive" (MSFD), the environmental pillar of the integrated maritime policy[20].

Consistency in measures for the conservation of marine biological resources and preservation measures adopted as part of environmental policy, such as protected marine areas, is therefore vital, particularly in vulnerable zones that are especially rich in marine biodiversity, such as coastal zones. Note that basic regulation (EU) no. 1380/2013 authorizes member states to adopt resource conservation measures as necessary in order

17 See today article 11 of the TFEU.
18 Article 2 of (EU) regulation no. 1380/2013, cited above. The precautionary approach in matters of fishery management is taken from article 191, section 2, first paragraph, of the TFEU and is defined as "the approach by which the lack of pertinent scientific data should not be used as a reason for postponing or failing to take management measures to conserve target species, associated or dependent species and nontarget species and their environment": article 4-1-8) of (EU) regulation 1380/2013, cited above. Ecosystem-based approach to fisheries management means: "an integrated approach to managing fisheries within ecologically meaningful boundaries which seeks to manage the use of natural resources, taking account of fishing and other human activities, while preserving both the biological wealth and the biological processes necessary to safeguard the composition, structure and functioning of the habitats of the ecosystem affected, by taking into account the knowledge and uncertainties regarding biotic, abiotic and human components of ecosystems". article 4-1-9) of (EU) regulation 1380/2013, cited above.
19 Regulation proposition by European Parliament and Council relative to common fisheries policy, Com (2011) 425 final, July 13, 2011, p. 6.
20 European Parliament and Council Directive 2008/56/CE of June 17, 2008 establishing a framework for community action in the field of marine environmental policy (Marine Strategy Framework Directive), OJEU no. L 164 of June 25, 2008, p. 19.

to comply with the requirements of the Union's environmental legislation[21], provided that these measures are compatible with the objectives of the CFP (article 11).

Before proceeding with a more in-depth analysis of these resource conservation measures and in order better to understand their impact, it is of primary importance to identify the delicate matter of decision-making within the EU itself.

4.1.1.3. *Who makes decisions within the European Union? A subtle sharing of competences*

In the earliest days of the European Community, the Council, made up of ministers from the various member states, was the sole decision-making body. As years went by, the European Parliament became a part of the decision-making process. The progressive development of recourse to a so-called co-decisional procedure enabled the newly-strengthened European Parliament, the only European institution elected by direct universal suffrage, to democratize the decision-making process, involving the Council and the European Parliament equally as it does (article 294, TFEU).

By virtue of this procedure, it falls to the European Commission to take the initiative by proposing legislation to the Council and the European Parliament. This is followed by various readings and back-and-forth discussions between these institutions, during which the draft document may be amended.

Ultimately, the text must be approved by both the European Parliament and the Council; if one of these two institutions refuses to approve the text, it will not be adopted.

Now called "ordinary legislative procedure", co-decisional procedure is the favored decision-making procedure within the EU, with the number of areas within which the Council remains the sole decision-making authority having shrunk over the years.

What does this mean for the domain of exploitation of marine biological resources?

---

21 Particularly concerned here is respect for the Habitat Directive (directive 92/43/CEE) and the Marine Strategy Framework Directive (directive 2008/56/CE).

It is important to recognize that the clarity of treaties on this subject leaves something to be desired. In accordance with article 43 of the TFEU, the Commission presents legislative proposals aimed at developing and implementing the CFP. Next, the European Parliament and the Council, ruling in accordance with ordinary legislative procedure, establish common market organization (CMO) in fishery products "as well as other provisions necessary for the pursuance of the objectives of the CFP" (article 43-2 TFEU).

The TFEU (article 43-3) specifies, however, that the Council remains the sole decision-making authority for the adoption of "measures on fixing prices, levies, aid and quantitative limitations and on the fixing and allocation of fishing opportunities".

The language of the treaty is not clear, and the wording of article 43-3 TFEU is not distinct enough from that of article 43-2 TFEU, particularly in the areas of markets and competition.

Finally, with regard to decision-making, we must not forget that the development and implementation of a decision involve a variety of committees and advisory boards, made up according to applicability of scientific experts[22], representatives of member states[23], or shareholders, fishermen, representatives of the processing and commercialization sectors, and representatives of environmental and consumer interest groups[24]. These committees and boards express opinions, in certain compulsory cases, which contribute to decision-making and illustrate the EU's wish to strengthen its mechanisms of participative democracy.

---

22 See the Comité scientifique, technique et économique de la pêche [scientific, technical, and economic committee for fisheries] (CSTEP): "The scientific, technical, and economic committee for fisheries, established by Commission decision 2005/629/CE, can be consulted on questions having to do with the conservation and management of marine biological resources in order to ensure the necessary participation of highly qualified scientists, notably in the biological, economic, environmental, social, and technological disciplines" considering no. 48 of (EU) regulation no. 1380/2013, cited above.
23 See the commission on fisheries and aquaculture, article 47 of (EU) regulation 1380/2013, cited above.
24 "Consulting boards have been established for each geographic zone (Baltic Sea, Black Sea, Mediterranean Sea, North Sea, northwestern waters and ultraperipheral waters) in order to foster a balanced representation of all stakeholders": see articles 43 and following of regulation 1380/2013, cited above.

Whether in matters of relations between the Union and its member states or of those between the various European institutions, this question of distribution of competence appears rather complex, and a source of potential conflicts.

However, it is on these foundations that the European policy on the conservation and management of living marine resources has been developed. Analysis will show how the Union and its member states are able to connect their respective actions in order to fulfill the objectives of this policy (see section 4.1.4), and we will then examine measures aimed more generally at supporting the CFP (see section 4.1.5).

4.1.2. Fundamental principles of common fisheries policy

The EU, which today is ranked fourth out of the world's fishing industries[25], is facing questions about the future of its CFP, based on principles and regulations adopted more than 30 years ago, the effectiveness of which sometimes leaves something to be desired. As noted by the European Commission in the introduction to its reflections on the CFP reform made effective on January 1, 2014, "Fish stocks are overfished, the economic situation of parts of the fleet is fragile despite receiving high levels of subsidies, jobs in the fishing sector are unattractive, and the situation of many coastal communities depending on fisheries is precarious[26]".

In order to understand this observation and to grasp the direction of the reforms, it is essential to examine the principal measures taken for the conservation of marine biological resources and the management of fisheries, as well as those taken to control them. First, though, it is important to emphasize the fact that this policy is based on one principle; that of equality of access to the waters and resources of the EU.

---

25 The EU's fishing catch production in 2011 was 6.143 million tons. The leading fish-catching country in the world is China (16.046 million tons), followed by Peru and then Indonesia. Source: Common fisheries policy in numbers, European Commission, basic statistical data, European Union Publications Office, 2014.

26 Communication from the Commission to the European Parliament, the Council, the European Economic and Social Committee, and the Committee of the Regions, "Reform of the Fisheries Common Policy" July 13, 2011, Com (2011) 417 final.

4.1.2.1. *Principle of equality of access to European Union waters and resources*

In accordance with the principle of non-discrimination on the basis of nationality (article 18, TFEU), all fishing vessels flying the flag of a EU member state enjoy access to the waters and resources of all EU waters[27]. However, there are exceptions to this principle.

In waters located less than 12 marine miles from baselines falling under their sovereignty or jurisdiction, member states are authorized to limit fishing rights to vessels traditionally operating in these waters out of ports on the adjacent coast. This important exemption from the principle of equality of access is justified in the first place by the desire to guarantee the economic stability of artisanal coastal fishing; it is also intended to reduce stress caused by fishery activities in maritime zones that are especially vulnerable from a biological point of view.

Given its exceptional character, this restriction of the equal-access principle is fixed for a limited period. In 1983, the date of the first regulation pertaining to the conservation of fishery resources, this derogation was set for a duration of 10 years. Since then, it has been renewed every 10 years. Regulation 1980/2013 repeats this schema; in waters situated less than 12 nautical miles from baselines falling under their sovereignty or jurisdiction, member states are authorized until December 31, 2022 to limit fishing rights to vessels traditionally operating in these waters out of ports on the adjacent coast (article 5-2). Thus, the principle of equality of access finds itself under attack, and some have declared it a "rampant nationalization of coastal zones" [LEB 11].

The possibility offered to member states to limit access for fishing vessels in areas within 12 miles of their coastlines was created with the particular aim of taking into account the neighborly relationships existing between them. For each member state, the basic regulation defines, in appendix I, the geographic zones of the coastal waters of the other member

27 Article 5 of (EU) regulation no. 1380/2013, cited above. European Union waters are defined as waters falling under the sovereignty or jurisdiction of its member states, with the exception of waters adjacent to overseas countries and territories (PTOM) connected to the European Union (these include New Caledonia and its dependencies, French Polynesia, the French Arctic and Antarctic territories, the Wallis-et-Futuna Islands, Mayotte and Saint-Pierre-et-Miquelon). Today, waters falling under the sovereignty or jurisdiction of member states may extend to up to 200 nautical miles beyond the baselines of member states.

states in which their ships may carry out fishing activities, as well as the species concerned[28].

Finally, as the last exception to the principle of equality of access, article 5-2 of (EU) regulation no. 1380/2013 states that, in waters situated less than 100 nautical miles from the baselines of the Union's ultraperipheral regions[29], the member states concerned are authorized, through December 31, 2022, to limit fishing to ships registered in the ports of these territories. These limitations do not apply to Union ships that traditionally fish in these waters, if these ships do not exceed customary fishing activity limits.

This new exception to the principle of equality of access is justified by the need to protect the vulnerable biological status of the waters around these islands, to take into account their structural, social and economic situations, and to preserve their local economies.

These exemptions from the principle of equality of access for the benefit of coastal EU member states enable the latter to adopt their own measures for the management and conservation of resources, provided that no measures have been adopted by the Union in these areas. These measures must also be non-discriminatory and compatible with the objectives of the CFP, and at least as strict as existing European regulations (article 20, (EU) regulation no. 1380/2013).

Member states may thus adapt European mechanisms imposed by analysis for the conservation of resources to fit specific regional characteristics.

4.1.2.2. *Conservation of marine biological resources*

There are two principal categories of resource conservation measures: quantitative rules aimed at managing resources by determining fishing possibilities that may be available to European fishermen (section 4.1.4.2.1), and qualitative rules, also called technical measures, which set conditions for

---

28 Fishing vessels flying the flags of the Netherlands, Belgium, the United Kingdom, Germany and Spain are authorized to fish in certain areas between 6 and 12 nautical miles off the coast of France. Fishing vessels flying the French flag are authorized to fish in the coastal areas off the coasts of Belgium (between 3 and 12 miles), Ireland (between 6 and 12 miles), the Netherlands (between 6 and 12 miles) and Spain (between 6 and 12 miles).
29 These include Guadeloupe, French Guyana, Martinique, Réunion, St. Barthélemy, St. Martin, the Azores, Madeira and the Canary Islands.

the use and structure of fishing gears, as well as restrictions of access to fishing zones (section 4.1.4.2.2).

4.1.2.2.1. Determination of fishing possibilities: strengthening multi-year management of resources

Managing fishing possibilities first and foremost involves the setting of catch amounts so as to comply with the objectives of the CFP.

Traditionally, fishing possibilities have taken the form of total allowable catches (TACs); that is the maximum quantities of fish that can be taken in and landed for each stock in a given zone. These TACs are then divided among member states by means of national quotas so as to guarantee each member state stability with regard to fishing activities for each fish stock or fishery (article 16, (EU) regulation no. 1380/2013). This stability must take into account the particular needs of regions in which the local communities are particularly dependent on fishing and related activities.

The rule of relative stability, which is a key to the distribution of catches, consists of allocating to each member state a fixed percentage of available catch possibilities. Today, considered to be the "central pillar" of the CFP[30], it constitutes a true principle of distribution of fishing opportunities, which guides the activities conducted within the EU throughout the year[31], leading to the adoption of TACs and quotas at the end of each year.

Through the years, the EU has improved the TAC and quota system with a view to strengthening the long-term visibility of fishery management[32].

Thus, the Council may adopt multi-annual fishery management plans. These plans, which "are key to conservation[33]", set forth management

---

30 European Parliament and Council legislative proposal relative to common fisheries policy, July 13, 2011, Com (2011) 425 final, p. 4.
31 Before submitting proposals to the Council, the European Union seeks the recommendations of various councils and committees, in particular the scientific, technical and economic committee for fisheries.
32 The setting of TACs must also incorporate environmental preservation requirements: so-called precautionary TACs must be set for stocks for which there is insufficient or unreliable data used to establish estimates of abundance.
33 COM (2011) 425 final, cited above, p. 7.

measures aimed at re-establishing or maintaining fish stocks above the levels necessary to obtain maximum sustainable yield[34].

Multi-year plans cover either fisheries exploiting single halieutic stocks, or fisheries exploiting multiple stocks in a determined geographic zone (mixed fisheries or interdependent stocks). In this case, they must include interactions between stocks and fisheries. These plans define objectives expressed in mortality rate per catch and in biomass of spawning stock, organize technical measures and set clear deadlines to be met.

Besides the measures adopted at the European level, each member state can develop conservation measures that are compatible with CFP objectives and applicable only to ships flying its flag or to individuals settled on their territory (article 19 of (EU) regulation 1380/2013). These measures, which must be at least as strict as existing European measures, must also be brought to the attention of other member states.

4.1.2.2.2. Technical measures

Technical measures are defined as "measures that regulate the composition of catches by species and size and the impacts on components of the ecosystems resulting from fishing activities by establishing conditions for the use and structure of fishing gear and restrictions on access to fishing areas" (article 4-20, (EU) regulation no. 1380/2013).

These measures include minimum landing size, compulsory use of selective fishing gears, setting of no-fishing periods in determined zones[35] and measures aimed at protecting the marine environment[36], in accordance with the environmental consistency principle.

---

34 Article 9, (EU) regulation no. 1380/2013. Maximum sustainable yield (MSY) means the highest theoretical equilibrium yield that can be continuously taken on average from a stock under existing average environmental conditions without significantly affecting the reproduction process (article 4-7 of (EU) regulation no. 1380/2013).

35 The key regulation establishing these various technical measures is (EC) Council regulation 850/98 of March 30, 1998, for the conservation of fishery resources through technical measures for the protection of juveniles of marine organisms, OJEC no. L 125 of April 27, 1998, p. 1 (regulation modified many times).

36 Here, we may cite Council regulation (EC) 1185/2003 of June 26, 2003 on the removal of fins of sharks on board vessels, OJEU no. L 167 of July 4, 2003, p. 1, as well as Council regulation (EC) 812/2004 of April 26, 2004 laying down measures concerning incidental catches of cetaceans in fisheries and amending Regulation (EC) No 88/98, OJEU no. L 185 of May 24, 2004, p. 4.

Finally, the EU is now seeking to limit by-catches dramatically; that is accidental, unintended catches of non-targeted species, by taking anti-discarding measures. Discards are a source of unacceptable waste both economically and environmentally; between 20 and 60% of fished species are discarded at sea[37]. In order to reduce the occurrence of this type of phenomenon, the CFP reforms made effective on January 1, 2014 institute the progressive requirement to land all catches[38].

As part of the technical measures established by the Union, member states may act by notifying the Commission and other member states of measures taken. This notification enables the Commission to assess both the compatibility of national measures with European regulations and their effectiveness.

If states do not inform the Commission of national measures taken, or if these national measures are not compatible with the objectives defined by the framework, or prevent the achievement of these objectives, the Commission may adopt measures by default.

With resource conservation measures being set, the sustainable management of fishing possibilities subsequently requires the adaptation of fishing vessels (in terms of size and engine power) and their activities' effects on the status of the resource.

4.1.2.3. Management of fishing capacity

Fishing capacity is defined as "a vessel's tonnage in GT (Gross Tonnage) and its power in kW (Kilowatt)[39]". Logically, member states must adapt their fleets' capacity to the fishing possibilities allocated to them in order to avoid any incidence of overcapacity that would be harmful to the sustainable

---

37 See *European Commission – Studies in the field of the Common fisheries policy and Maritime affairs – Impact assessment of discard reducing policies*, June 2011. The percentage of disposals is liable to vary greatly depending on the fishing technique used.
38 Article 15, (EU) regulation no. 1380/2014. Mandatory landing will be progressively implemented between 2015 and 2019. Moreover, limited acceptable disposals (5%) are possible in order to aid the fishing industry to adapt to the requirement of landing all catches. *"All catches of species which are subject to catch limits [...] caught during fishing activities in Union waters or by Union fishing vessels outside Union waters in waters not subject to third countries' sovereignty or jurisdiction, in the fisheries and geographical areas listed blow shall be brought and retained on board the fishing vessels, recorded, landed and counted against the quotas [...]"*
39 Article 4-24, (EU) regulation no. 1380/2013.

management of the resource. This requires member states to monitor closely the evolution of their fishing fleets and to transmit a report each year to the Commission assessing their fishing capacity. If this assessment shows overcapacity, the state must adopt a plan of action aimed at balancing fishing capacity of its fleet with fishing possibilities. Member states must also manage entries into and exits from the fleet "in such a way that the entry into the fleet of new capacity without public aid is compensated for by the prior withdrawal of capacity without public aid of at least the same amount" (article 23, (EU) regulation no. 1380/2013).

Fishing capacity management is also conducted by controlling fishing efforts, which are defined for a ship as "the product of its activity and its capacity[40]".

Managing fishing activity makes it necessary to have a precise knowledge of the state of the European fishing fleet. For this reason, European legislation requires all fishing vessels to hold fishing licenses. Fishing efforts can thus be limited by the conferral of special fishing licenses to a limited number of vessels, authorized to fish for certain specific resources in a determined zone. It is also possible to limit fishing activity by imposing periods in which fishing is banned (which may be compensated financially).

In reality, the management of fishing efforts seems quite complex to implement, due in particular to the difficulties of controlling both vessels' capacity (especially the use of fishing gear) and their effective activity. The Commission noted in 2013 that "the capacity of the Europe's fishing fleet is still too high [...] The current fleet management policy has failed to bring fleets into balance with the resources they exploit[41]".

This observation had, in 2011, led the Commission to propose the granting to shipowners of transferable fishing concessions defined as "revocable user entitlement to a specific part of fishing opportunities allocated to a Member State or established in a management plan adopted by a Member State [...] which the holder may transfer to other eligible holders

---

40 Article 4-21, (EU) regulation no. 1380/2013. For a group of fishing vessels, fishing efforts are defined as "the sum of the fishing efforts of all the vessels in the group". Capacity can be measured according to the size of ships (tonnage), the fishing gears used or their engine power (KW). Activity is generally measured by the number of days spent at sea.
41 Commission report to the European Parliament and Council relative to efforts deployed by member states in 2011 to achieve a sustainable balance between fishing capacity and fishing possibilities, COM (2013) 85 final, February 18, 2013, p. 11.

of such transferable fishing concessions[42]". In this, the Commission perceived an effective means of fighting against fishing overcapacity, a fundamental problem of the CFP[43]. According to it, these transferable fishing rights would constitute a more efficient and less expensive way of remedying the overcapacity of fleets, while increasing responsibility in the sector[44].

Ultimately, regulation (EU) 1380/2013 included transferable fishery concessions[45] among its fishing capacity management measures. However, states remain free to institute this type of system of transferable individual quotas, if they so desire[46]. The Union opted not to impose this system, which brought up significant fears regarding the concentration of fishing rights ownership and socioeconomic imbalance.

Now that the major outlines of the European system for resource conservation and fisheries management have been sketched, it is important to examine the mechanisms put in place by the EU to guarantee compliance with this system.

4.1.2.4. *The delicate matter of ensuring compliance with regulations*

Ensuring compliance with regulations is a major element of the marine biological resources conservation system[47]. The adoption of norms must be accompanied by the certainty that they will be respected. Moreover, ensuring

---

42 Com (2011) 425 final, cited above.

43 Communication from the Commission to the European Parliament, the Council, the European Economic and Social Commission, and the Committee of the Regions, "Reform of the Fisheries Common Policy", Com (2011) 417 final, July 13, 2011, p. 5.

44 In reality, the holders of these individual rights would no longer be motivated to increase their fishing capacity in order to access the resource, in accordance with the so-called "Tragedy of the Commons", by virtue of which free access to a limited resource results in the overexploitation of this resource and eventually causes its disappearance. On this point, see [HAR 68].

45 "transferable fishing concession" means a revocable user entitlement to a specific part of fishing opportunities allocated to a Member State or established in a management plan adopted by a Member State in accordance with Article 19 of Council Regulation (EC) No 1967/2006 (1), which the holder may transfer" (article 4-23 (EU) regulation no. 1380/2013, cited above).

46 See article 21 ((EU) regulation no. 1380/2013). Some Union member states already use this type of system, including Denmark, Spain, the Netherlands and the United Kingdom.

47 Article 36 ((EU) regulation no. 1380/2013): "Compliance with the CFP rules shall be ensured through an effective Union fisheries control system, including the fight against IUU fishing". This control relies on cooperation and coordination between member states, the Commission and the European Fisheries Control Agency.

compliance with regulations in the domain of marine biological resource conservation is particularly delicate in view of the environment in which the activity is conducted – the marine environment – and the large number of texts dedicated to the subject.

Regulation (EC) no. 1224/2009[48], called the "conformity regulation", is the fruit of years of reflection begun in 1982[49] with the adoption of the first fishery control mechanisms. This text established a community-wide system of controls, inspection and execution in order to ensure compliance with all rules and regulations of the CFP[50] (article 1). It does not, then, limit itself solely to ensuring compliance with resource conservation regulations for community vessels[51].

The European system of control emphasizes the primary responsibility of member states, which are tasked with controlling activities conducted as part of the CFP on their territory or in waters falling under their sovereignty or jurisdiction (article 5). It also falls to member states to control access to waters and resources as well as activities conducted outside community waters by vessels flying their flag.

The framework set by (EC) regulation no. 1224/2009 relies principally on a system of licensing. A fishing vessel cannot exercise its activities unless it holds a fishing license issued by the flag state (article 6). This system enables member states to manage their fishing fleets in order to ensure that

---

48 (EC) Council regulation no. 1224/2009 of November 20, 2009 establishing a Community control system for ensuring compliance with the rules of the common fisheries policy, amending Regulations (EC) No 847/96, (EC) No 2371/2002, (EC) No 811/2004, (EC) No 768/2005, (EC) No 2115/2005, (EC) No 2166/2005, (EC) No 388/2006, (EC) No 509/2007, (EC) No 676/2007, (EC) No 1098/2007, (EC) No 1300/2008, (EC) No 1342/2008 and repealing Regulations (EEC) No 2847/93, (EC) No 1627/94 and (EC) No 1966/2006, OJEU no. L 349 of December 22, 2009, p. 1.

49 (EEC) Council regulation no. 2057/82 of June 29, 1982 establishing certain control measures with regard to fishery activities conducted by member state vessels, OJEC no. L 220 of July 29, 1982, p. 1.

50 (EC) regulation no. 1224/2009 also institutes a control on recreational fishing activities that may have a significant impact on fishery resources. Recreational fishing activities are defined as "non-commercial fishing activities exploiting living marine aquatic resources for recreation, tourism or sport": article 4-28 of regulation (EC) 1224/2009. Member states must ensure that these activities are carried out in ways compatible with the objectives of the common fisheries policy.

51 See later the control system in other areas of the CFP, particularly in matters of the commercialization of resources.

the capacity corresponding to fishing licenses is not greater than the maximum capacity levels attributed to them.

Moreover, fishing authorization is required for any ship conducting a strictly regulated activity, particularly in the context of a system to manage fishing efforts, a multi-year plan or falling within a restricted fishing zone (article 7).

Once at sea, the fishing vessel is subject to specific surveillance; fishing ships flying the flag of a member state, like fishing vessels from third-party countries practicing their activities in EU waters, must be equipped with a vessel monitoring system[52].

(EC) regulation 1224/2009 confers particular responsibility on the masters of fishing vessels[53], who are required to keep a fishery logbook of their activities (dates of catches, quantities fished, etc., must be recorded), and to declare transshipments of catches as well as landings carried out.

The monitoring of fishing efforts is also the subject of a specific control, in view of the necessity of having precise and accurate knowledge of the activities of the vessel concerned, a basic element used to calculate fishing efforts; member states must strictly control the presence of ships in fishing zones as well as the use of the fishing equipment concerned.

More broadly, each member state must monitor the evolution of fisheries and prohibit the fishing of a species once the quota attributed to it for that species has been reached. If the Commission sees that fishing possibilities have been exhausted without the appropriate reaction from the state, it falls upon the Commission to prohibit this fishing. It also falls to member states to ensure compliance with technical measures, in terms of fishing equipment used as well as fishing periods allowed.

Member states are thus at the heart of the European system for ensuring compliance with measures to conserve marine biological resources. This is a weighty responsibility, and one sometimes unwillingly assumed, as shown

---

52 (EC) regulation 1224/2009, however, specifies exceptions, notably for vessels less than 15 m in length operating exclusively in the territorial waters of the member state whose flag they fly.
53 This includes fishing vessels longer than 10 m overall.

by the condemnation of France by the European Court of Justice to heavy financial sanctions for failure to fulfill its obligations in matters of control[54].

Control systems put in place by member states are themselves subject to checks by community inspectors who can carry out inspections with or without national agents present, on board ships or on the premises of fishery businesses.

Community inspectors are not endowed with police powers. Following their inspection missions, they write reports which, once transmitted to member states, enable these member states to improve their control systems if flaws have been observed[55].

Management of fishery possibilities, technical measures and the ensuring of compliance with regulations constitute the three major wings of the European system for the conservation of marine biological resources. The sustainable management and exploitation of these resources also requires an appropriate economic and structural environment that guarantees the effectiveness of this system.

4.1.3. *Definition of an economic framework for sustainable exploitation of marine biological resources*

The marine biological resource conservation system cannot work at maximum effectiveness if it is not part of an appropriate economic framework, both within Europe (section 4.1.5.1) and outside it, given the necessity for European fishers to operate outside European waters (section 4.1.5.2).

---

54 CJCE July 12, 2005, Commission C./France, aff. C-304/02, Rec. 2005- I-6263. For, in particular, not having ensured compliance with minimum fish size, France was required to pay the Commission damages in the amount of 57,761,250 euros for each 6-month period counting from the handing-down of the decree and until it complied with its obligations. It was also required to pay the Commission a fixed sum of 20,000,000 euros.

55 In addition to the European body of inspectors, the European control system was enforced by the 2005 creation of the community fisheries control agency; see (EC) Council regulation no. 768/2005 of April 26, 2005 instituting a community fisheries control agency, JOCE no. L 128 of May 21, 2005, p. 1. The mission of this agency is to coordinate national control agencies and to assist them in cooperating in order to guarantee the effective and uniform application of CFP regulations (article 1 of (EC) regulation 768/2005).

4.1.3.1. *An intra-European economic framework for the sustainable exploitation of marine biological resources*

The CFP covers not only the conservation of marine biological resources, but also measures[56] concerning the organization of markets for fishery products on one hand (section 4.1.5.1.1) and financial instruments aimed at aiding both member states and operators on the other hand[57] (section 4.1.5.1.2).

4.1.3.1.1. Common organization of the market in fishery products

This organization attests to the importance placed by the EU on the market for fishery products, as does the fact that the first measures relative to the CFP were adopted on this subject in 1970[58]. The common organization of the market in fishery products (henceforth referred to as the CMO) has been reformed several times, most recently with the adoption of (EU) European Parliament and Council regulation no. 1379/2013[59].

The goal of the CMO is, above all, to strengthen competitiveness in the fishery sector, in particular that of producers; to improve the transparency and stability of markets; to contribute to ensuring equal conditions for all products commercialized within the Union and also to guarantee consumers the availability of a diverse supply of fishery products while also providing them with accurate and verifiable information on the origin of the product[60].

The CMO is based on producers' organizations (POs), which constitute "keys to achieving the objectives of the common fisheries policy and the common market organization[61]". POs are formed on the initiative of producers of fishery products. Upon their request, they can be recognized by

---

56 See article 1 of (EU) regulation 1380/2013, cited above.

57 An operator is defined as any individual or corporation managing or owning a business that exercises an activity connected to any stage of the production, processing, commercialization, distribution and retail sales chains of fishery and aquaculture products: article 4-10, regulation 1380/2013, cited above.

58 (EEC) Council regulation no. 2142/70 of October 20, 1970 on the common organization of the market in the fishery products sector, JOCE no. L 236 of October 27, 1970, p. 5.

59 (EC) European Parliament and Counci regulation no. 1379/2013 of December 11, 2013 on the common organization of the market in the fishery and aquaculture products sector, modifying Council regulations (EC) no. 1184/2006 and (EC) no. 1224/2009 and abrogating (EC) Council regulation no. 104/2000, OJEU no. L 354 of December 28, 2013, p. 1.

60 Article 35, regulation 1380/2013, cited above.

61 In view of no. 7 of (EU) regulation no. 1379/2013, cited above.

member states if they fulfill certain *criteria* such as the exercise of adequate economic activity (though without holding a dominant position on the market).

The goal of POs is to promote the exercise by their members of viable and sustainable fishing activities, "in full compliance with the conservation policy, as laid down, in particular, in Regulation (EU) No 1380/2013 and in environmental law, while respecting social policy and, where the Member State concerned so provides, participating in the management of marine biological resources[62]".

POs must take responsibility for unintended by-catches included in commercial stocks; to do this, they may plan the fishing activities of their members, put landed products on the market and adapt production to market requirements[63].

POs play a decisive role in the implementation of the CFP as, in addition to their actions on the market, they must manage the subquotas allocated to them by member states. The POs must submit to national authorities production and commercialization plans that are then approved by these authorities. The missions assigned to POs by regulation (EU) no. 1379/20133 thus greatly exceed simple market organization; they are truly at the interface of resource management and market organization[64].

In view of this vital role played by POs, European competition regulations prohibiting accords between businesses preventing, limiting or misrepresenting competition are not applicable to the actions of POs having to do with the production or sale of fishery products or the use of common storage, treatment or processing facilities for these products, particularly if these actions are necessary for the achievement of CFP objectives (article 41 of (EU) regulation no. 1379/2013). As we can see, POs clearly constitute specific governance actors within the EU.

---

62 Article 7 of (EU) regulation no. 1379/2013, cited above.
63 In order to stabilize the markets, POs may finance the storage of a certain number of fishery products if these products have not found buyers on the market: articles 30 and following of (EU) regulation no. 1379/2013, cited above.
64 The position of POs is all the stronger because, while membership in POs is optional, member states may make compulsory the regulations they adopt for producers which are not members: article 22, (EU) regulation no. 1379/2013, cited above.

Another major aspect of the CMO consists of the setting of common commercialization standards[65]. These standards notably set minimum sizes for commercialization. The products concerned cannot be commercialized for human consumption within the EU if they do not comply with these standards. The institution of common commercialization standards in 1970 has contributed greatly to greater market transparency and has fostered trade in fishing products between member states.

Another important point for the CMO is consumer information.

Observing that "the potential of the EU market remains largely unexploited, and the increase in consumption throughout the EU offers real economic possibilities for its producers", the CMO takes a keen interest in consumers, particularly by seeking to give them highly accurate information to strengthen their trust in fishery products[66]. This information appears more necessary than ever, at a time of increasing diversity of supply. Additionally, fishery products can only be put on the market if they possess appropriate display and labeling[67].

Finally, it is important to note that the commercialization of marine products is subject to strict controls as described in (EC) regulation 1224/2009, which institutes a community-wide control system[68]. In order to ensure the traceability of products and thus to ensure consumer confidence, quantities unloaded, put up for sale and purchased must be declared. Member states must ensure compliance with regulations instituted by the CMO, such as common standards of commercialization. These are intended to ensure that all fishery products are commercialized or registered in an auction house or with approved purchasers or POs[69].

Given their significant responsibility in terms of putting products on the market, POs are closely controlled: a member state which observes that a PO is not complying with the regulations of the CMO must immediately withdraw its formal recognition. Finally, member states are responsible for

---

65 See Chapter 3 (articles 33 and following) of (EU) regulation no. 1379/2013, cited above.
66 See Chapter 4 (articles 35 and following) of (EU) regulation no. 1379/2013, cited above.
67 Display or labeling must indicate the commercial name of the species, the method of production, the area and date of catch, and whether the product is fresh or has been frozen and thawed: article 35 of regulation (EU) no. 1379/2013 cited above.
68 (EC) Council regulation no. 1224/2009 of November 20, 2009 cited above.
69 However, a purchaser who acquires, for a maximum weight of 30 kg, fishery products that are not subsequently placed on the market but used uniquely for private ends, is exempt from this obligation (article 59, (EC) regulation no. 1224/2009, cited above).

ensuring compliance with prices and intervention systems (withdrawal of products from the market, private storage, etc.).

In addition to the CMO, financial instruments intended to aid member states and operators form a major wing of the CFP.

4.1.3.1.2. Financial instruments

In order to fulfill the objectives of the CFP, the EU may allocate financial aid to member states and operators. This financial contribution falls within the EU structural policy aimed at ensuring economic, social and territorial cohesion within the Union, in accordance with articles 174 and following of the TFEU.

In this context, the Union seeks to reduce the discrepancy between levels of development in various regions, as well as the lagging behind of less favored regions (article 174 pgh.2 TFEU). To achieve these ambitious goals, the EU uses various financial instruments that take the form of European funds[70] with complementary objectives. With regard to fishery, the CFP reform made effective on January 1, 2014 provides for the establishment of a dedicated new fund, the European Maritime and Fisheries Fund (EMFF)[71]. The difficulties of member states in getting along with one another on this delicate financial question explain the delay in adopting this fund, with political agreement between the European Parliament and the Council having only occurred on January 25, 2014.

The EMFF[72], the objective of which goes beyond the simple context of fishing as it must also comply with the objectives of integrated maritime

---

70 These funds include the European Regional Development Fund (ERDF), the European Social Fund (ESF), the Cohesion Fund and the European Agricultural Fund for Regional Development (EAFRD).

71 European Parliament and Council (EU) regulation no. 508/2014 of May 15, 2014 relative to the EMFF and abrogating Council regulations (CE) no. 2328/2003, (CE) no. 861/2006, (CE) no. 1198/2006 and (CE) no. 791/2007 and European Parliament and Council regulation (EU) no. 1244/2011, OJEU no. L 149 of May 20, 2014, p. 1. The EMFF covers the period 2014–2020 and succeeds the European Fisheries Fund (EFF) which, endowed with 4.3 billion euros, covered the period 2007–2013.

72 The EMFF is endowed for the period 2014–2020 with 5,749,331,600 euros in shared management (4,340,800,000 euros of which are allocated to the sustainable development of fishery, aquaculture and areas dependent on fishing, and to measures having to do with commercialization and processing and with technical assistance on request of member states), and 647,275,400 euros in direct management. In the case of shared management, member states and not Commission services choose the beneficiaries and manage expenditures. In the case of direct management, it is the Commission that selects funding beneficiaries, pays the funds and monitors their activities.

policy, is a cofinancing instrument. It is used in tandem with national financing, with each member state required to develop an individual operational program implementing the priorities of the Union and specifies the details of expenditure of funds obtained. Each operational program is evaluated and approved by the Commission.

The EMFF must promote durable and competitive fishery and the balanced and united territorial development of areas dependent on fishery, in accordance with the economic and social consistency policy, and foster the implementation of the CFP.

To do this, financial aids must contribute to the sustainable development of fishery by limiting fishing activity in the marine environment and by protecting marine biodiversity.

In this context, some operations cannot hope to obtain EMFF support: operations increasing the fishing capacity of a vessel; the construction of new fishing vessels; experimental fishery, etc.

Conversely, the EMFF favors the use of more selective fishing gears which reduces by-catches and discarding, and supports shipboard investments aimed at adding value to the underused part of catches.

It also takes an interest in shipboard working conditions by supporting investments in individual equipment that exceeds the standards imposed by national law or the EU.

The EMFF also helps to protect and re-establish marine biodiversity and ecosystems, for example by fostering the collection of sea waste such as lost fishing gears, and by supporting the surveillance of protected marine areas.

On land, the EMFF supports landing sites that foster energy efficiency and the quality of landed products, safety and work conditions.

Note that the EMFF's interests do not lie only in production activities; rather, its actions involve the entire fisheries industry. It contributes to the financing of activities of commercialization (storage assistance and help with promotional campaigns, etc.) and processing, which add value to fishery products.

The EMFF may also help to diversify the local economy in areas dependent on fishing in order to create growth and new jobs. It fosters the creation of businesses outside the fishing industry as well as the redevelopment of small coastal fishing vessels for their reassignment to activities other than fishing. In this case, the EMFF's actions must be coordinated with those of other EU funds, such as the European Regional Development Fund (ERDF) and the European Social Fund (ESF).

Finally, beyond the CFP alone but in close collaboration with it, the EMFF supports new tools that help to create synergy between different sectors related to sea industry, such as integrated maritime surveillance.

The CMO and financial instruments manifest the unilateral actions of the EU and have an essential intra-European scope. The EU has also developed a whole network of international partnerships aimed at instituting an economic framework that is equally favorable to the sustainable exploitation of marine biological resources.

4.1.3.2. *International openness: a guarantee for the survival of European fishery*

European fishermen have always conducted their activities outside waters falling under the sovereignty or jurisdiction of EU member states (called EU waters). Against a background of structural deficit in fishery products[73], the EU prioritizes two categories of actions: bilateral agreements (section 4.1.5.2.1) and multilateral relations (section 4.1.5.2.2).

4.1.3.2.1. Bilateral fishery agreements

In the first place, the EU has concluded agreements with the countries of northern Europe (Norway, Iceland, etc.) organizing the joint management of stocks situated in the North Sea and the northeast Atlantic. The joint management of these stocks is required when these stocks are located

---

73 In 2012, the European Union imported 19.2 million euros' worth (5.5 million tons) of fishery and aquaculture products, and exported 4.1 million euros' worth (1.9 million tons). Source: "Figures on the Common Fishery Policy", Eurostat, 2014.

simultaneously on the high seas and in waters under the jurisdiction of the Union member states and the Nordic states concerned. These agreements are crucial, as they involve large quantities of fishery resources[74].

In addition to these "Nordic" agreements, the EU has created a whole network of bilateral agreements called partnership agreements, or Sustainable Fisheries partnership Agreements (SFAs). As part of these SFAs, the EU pays financial compensation to third-party countries in exchange for access for European fishers to the resources present in the waters of these partner countries. These fisheries agreements are highly detailed, specifying the species and quantities able to be fished, the fishing gear to be used, and the type and number of vessels authorized to access the fishing zone concerned.

In order to aid the partner country, most often a developing country, to institute a sustainable fishery system, part of the financial compensation paid by the EU is designated for the promotion of this activity. European financial aid may consist of monitoring actions or scientific research, by means of which the EU hopes to contribute to a better conservation of marine biological resources outside European waters.

In addition to these bilateral fisheries agreements, the EU supports the conservation of marine biological resources outside the waters of its member states by participating in the international network of regional fisheries organizations.

4.1.3.2.2. The European Union as a member of regional fisheries organizations

A regional fisheries management organization (RFMO) is an intergovernmental organization created to manage one or more fish species in a determined maritime zone, most often on the high seas.

---

74 For 2014, see, for example, Council regulation (EU) no. 43/2014 of January 20, 2014 establishing, for 2014, fishery possibilities for certain fish stocks and groups of fish stocks, applicable in EU waters and, for Union ships, in certain waters not belonging to the EU, JOUE no. L 24 of January 28, 2014, p. 1. This regulation determines the distribution of fishery possibilities between the European Union, Iceland and Norway.

The EU is a member of approximately 50 RFMOs[75] and participates in the development of texts defining quantities that can be fished, periods when fishery is prohibited and vessel inspection measures.

RFMOs have above all been deeply involved, in recent years, in the fight against illegal, undeclared and unregulated fishery (called IUU fishery). Most of these RFMOs publish a list of vessels that practice this kind of fishery, and their members are requested to prohibit the landing or importation of catches resulting from this IUU fishery.

In complement to these actions taken by RFMOs, the EU adopted a regulation in 2008 aimed specifically at fighting this scourge by prohibiting the importation into the EU of fishery products resulting from IUU fishery[76].

Thus, RFMOs constitute for the EU a perfectly adequate framework allowing it to contribute to the sustainable management of marine biological resources throughout the world's oceans.

Thus, the EU's action continues beyond the borders of its member states, as required for the sustainable exploitation of marine biological resources. Its action on the international scene emphasizes the interest for its 28 member states of acting together in order to have a greater impact.

4.2. French policy on sustainable exploitation of marine living resources

In the fields of fishery and aquaculture, France occupies an estimable position within the EU[77]. Beyond the raw numbers, however, the current

---

75 These include the International Commission for the Conservation of Atlantic Tuna (ICCAT), the Northeast Atlantic Fisheries Commission (NEAFC), the Northwest Atlantic Fisheries Organization (NAFO), etc.

76 Article 12 of Council regulation (CE) no. 1005/2008 of September 29, 2008 establishing a community-wide system intended to prevent, discourage and eradicate IUU fishery, modifying regulations (EEC) no. 2847/93, (EC) no. 1936/2001, and (EC) no. 601/2004 and abrogating regulations (EC) no. 1093/94 and (EC) no. 1447/1999, JOUE no. L 286 of October 29, 2008, p. 1.

77 In 2010, the entire French fishing and aquaculture industry represented a total value of 1.7 billion euros (1 billion of which resulted from fishing and 700 million from aquaculture) for a production of 710,000 tons (484,000 tons in fishing and 236,000 tons in aquaculture); 7,300 vessels (4,860 of which operate out of metropolitan France), and a cumulative engine power of almost 1,000,000 KW. Finally, the industry employed 22,640 fishermen (585 of which were not native to the EU) and 20,000 employees in aquaculture (11,400 of which were full-time employees).

balance of activities is largely negative both in terms of the sector's economy and the sustainable conservation of resources[78]. In the face of this seemingly inevitable decline, French institutions have not remained inactive. Most of the initiatives taken fall within the general context of EU policy, particularly that of the CFP[79].

While the EU generally supervises the national system of management and conservation of fish resources, it does not specify all of its elements alone. In reality, the Union's competence extends over only a determined part of France's national waters, over which the state retains significant powers of action. Thus, there is indeed a French policy on the sustainable exploitation of fish resources. Before looking at the key instruments, it is advisable to examine the fundamental principles of this policy".

4.2.1. *Fundamental principles of French policy*

These fundamental principles arise from several areas and concern the delineation of competences, the affirmation of principles, the determination of objectives and the construction of a specific institutional framework.

4.2.1.1. *Competences*

France's competences in matters of the management and conservation of fish resources vary according to the system of the areas in which these resources are accessible to it. Without infringing upon the application of the provisions of international fisheries law to which France is subject, this competence varies according to whether it is in waters under sovereignty or jurisdiction that are covered or not covered by EU regulations.

In the first case, in waters bordering the territory of metropolitan France and overseas regions (French Guyana, Martinique, Guadeloupe, Réunion, Mayotte and St. Martin), France, as a member of the EU, exercises only delegated powers. In the second case, waters bordering all overseas

---

78 See, for example, [CLE 08] and [GUE 11].
79 See Regulation (EU) n° 1380/2013, above-mentioned. See also the French memorandum relative to CFP reform, ministry of Food, Agriculture and Fisheries, January 2010, and "Réponse des Autorités françaises au Livre Vert sur la politique maritime de l'Union", April 2007.

territories as well as Clipperton Island are subject to a special system of association defined in part four of the TFEU, for which France has full and complete competence subject to the implementation of constitutional provisions by virtue of which it delegates to these territories the exercise of some of its competences. With regard to the first case, remember that the exclusive competence held by the Union in the matter does not deprive the state of all its powers or prevent the normal exercise of its powers of execution. With regard to the second case, and depending on the case, these are subject to the principles of legislative identity or specialization.

4.2.1.2. Principles

The fisheries sector in France obeys basic principles whose application is eminently precautionary[80]. These concern the conditions of conducting fishing activities on one hand, and the status of fish resources on the other hand. The question of the nature of fishing activities can be connected to this whole.

Paradoxically, the exercise of fishery activities remains inextricably linked to the double principle of liberty and equality proclaimed by the royal ordinance of August 1681, in which "we declare fishing in the seas free and common to all those of our subjects who are granted permission to do so, both on the open sea and on the shores, with nets and gears allowed by this ordinance". Analysis of the facts, however, seriously undercuts the impact of each of the terms of this proclamation.

While it is true that the principle of freedom obviously does not infringe upon the fact that the right to fish can be regulated, it appears that the nature of the principle, as specified by the effective provisions of decrees involving the regulation of coastal fishing activities (four decrees from July 4, 1853 and one from November 19, 1859) – "fishing is maritime; that is, without rent or license" has been progressively drained of its substance.

First, barring exceptions, professional fishing activities are now subject to a system of prior authorization, whether in regard to the authorization to

---

80 The founding principles of fishing activity (freedom of exercise, equality and freedom from payment), though regularly reiterated, have today lost virtually all of their substance.

exercise the profession of fisherman in itself, or to authorization to collect this or that resource or to exercise this or that specific "occupation[81]".

Second, the principle of freedom from payment has also become largely obsolete, partly because the compulsory membership of fishers in a professional organization is accompanied by the payment of dues, and partly because it is common for the allocation of fishing authorizations to be accompanied by the payment of annual fees.

With regard to equality of access for French citizens to fish resources, this is manifestly an illusion, as commercial fishing activities are linked to the profession of fisher. Moreover, entry into this profession is subject to compulsory conditions such as the obtaining of a degree (with the first step being a certificate of aptitude to exercise the profession of fishing deckhand); membership in a social organization (the ENIM [*établissement national des invalides de la marine*, or welfare fund for seamen] and membership in a professional organization: national committee of maritime fishing and farming[82].

Fisheries resources enjoy a unique status. The law decrees that all fisheries resources to which France has access, "both in its waters under jurisdiction or sovereignty and in other waters in which it has fishing rights by virtue of international agreements or in high-seas areas" are part of a "collective heritage[83]". This qualification excludes any idea of a patrimonialization of fishery resources for private use, and thus any mechanism – for the appropriation and/or transferability of fishing rights – that might prove necessary for this purpose and makes the state the guarantor of the sustainable conservation of this heritage for the benefit of future generations. With regard to the priority access granted to professional fishermen, it seems that the collective patrimonialization of resources is part of the constitution of a system of "neo-ownership[84]" benefiting a limited part of this community.

The concept of a collective heritage is in sync, moreover, with the principle of excluding fishing boats flying the flag of a foreign state from

81 See infra section 4.2.2.
82 Article L.912-1 CRPM.
83 Article L.911-2, 1, *code rural et de la pêche maritime* [rural and maritime fisheries code] (CRPM).
84 To use an expression by Jean-Luc Prat [PRA 96].

fishing zones falling within waters under the sovereignty or jurisdiction of the state, subject to the provisions of EU law and international agreements made between the Union and France[85].

Finally, it is important to qualify the nature of fishery activities in legal terms.

Professional fishing is a commercial activity[86]. This acknowledgment involves the registration of skippers in the trade and companies register[87]. Moreover, the profits of their activities are taxable under the category of industrial and commercial revenue, except for income corresponding to shared wages earned by the skipper owners as part of their personal work, which are taxable under the category of pay and salary[88].

Leisure fishery is defined *a contrario* to the above. It is a fishing activity "of which the product is destined exclusively for consumption by the fisherman and his family and cannot be sold door-to-door, displayed for sale, or sold under any form whatsoever, or knowingly purchased. It is conducted either from on-board vessels or small boats other than those officially registered as fishing vessels, or during swimming or diving, or on foot in public maritime areas as well as from the banks of saltwater streams, rivers, or canals[89]".

Marine farming activities are considered agricultural, the social status of those who practice them notwithstanding[90].

4.2.1.3. Objectives

The objectives of French fishery policy, as given in the rural and maritime fishery code[91], were initially defined by the 1997 Outline Act on maritime fishery and marine farming "in accordance with the principles and regulations of the common fisheries policy and in compliance with international commitments". They are intended to provide for sustainable

85 Article L.921-9 CRPM.
86 Article L.931-1 CRPM.
87 Specific provisions are given for professional on-foot fishermen; see articles L.722-1, L.722-5 and D.722-5 CRPM.
88 See article 34 of the general tax code.
89 Decree no. 90-618 of July 11, 1990 relative to the exercise of leisure maritime fishery (article 1).
90 Article L.311-1 CRPM.
91 Article L.911-2.

exploitation, to add value to and adapt products of the fishing industry for the markets, to develop research, to sustain a policy of quality and traceability, to renew the fleet and modernize the fishing sector, to develop marine aquaculture and finally to diversify activities for the benefit of the economies of coastal regions.

If we set aside the objective of developing marine aquaculture, with regard to which the law makes express reference to environmental quality, we can see that, curiously, these objectives do not directly take into account the necessity of including maritime fishery in current imperatives relative to the protection and preservation of the marine environment[92].

However, other provisions of domestic law[93] make up in part for this "oversight". Stresses caused by human activities, particularly those of fishery and aquaculture, must generally be compatible with the achievement of the "good ecological state" of the marine environment. Indeed, the protection of this environment, the conservation of its biodiversity and its sustainable use by maritime and coastal activities while respecting marine habitats and ecosystems are "of general interest[94]". More specifically, fishery management policy must take into account provisions relative to "policies for the marine environment" in the environmental code[95] intended to transpose various provisions of the European directive entitled "strategy for the marine environment[96]".

It is also indicated that maritime fishing activities are subject to various provisions of the environmental code[97]. Upon analysis, this has to do more particularly with provisions relative to the "conservation of natural habitats of wild fauna and flora[98]" which are manifested by limitations imposed on fishing activities as part of the implementation of a system of marine protected areas (MPA) under the terms of the "habitats" directive[99]". Specifically, when the conservation of habitats and species on sites in the

---

92 Contrary to the objectives of the "new" CFP.
93 To which new provisions will undoubtedly be added, taken from future laws relative to biodiversity.
94 Article L.219-7 of the environmental code.
95 Section 1, book II, chapter IX.
96 Directive 2008/56/EC, above-mentioned.
97 Article L.921-10, 2nd paragraph, CRPM.
98 See Chapter 4 of book IV of section 1 of the environmental code.
99 Council Directive 92/43/EEC of 21 May 1992 on the conservation of natural habitats and of wild fauna and flora, OJEC n° L 206 of July 22, 1992.

"Natura 2000" network necessitates fishery management measures, the state concerned must notify the EU so that it can set the necessary regulations[100].

Finally, there are provisions specific to maritime fishery in certain marine protected areas such as national parks, wilderness areas, nature reserves and natural marine parks[101].

The pursuit of these objectives can no longer be seriously envisioned in a strictly sectorial context, given the fact that the development of maritime activities in their diversity tends to multiply interactions that only an integrated approach would be able to rationalize, and also because the conservation of fish stocks cannot take place without the use of an approach based on the management of ecosystems as a whole. Though these factors are neither new nor particularly original, their acknowledgment and integration have become an unavoidable part of the definition of public policies dedicated to maritime activities.

The ecosystemic approach[102] is undoubtedly the "new" paradigm of policies for the conservation and management of fish stocks. Even the lack of a direct reference to this concept in domestic fishing regulation does not mean it can be avoided, even in the application of imperative provisions of EU law, and more broadly in compliance with France's international commitments.

For now, the movement to integrate sectorial maritime policies is occurring mainly under cover of environmental legislation. Policies for so-called "integrated management of coastal areas" or "maritime spatial planning" are considered to constitute eventual driving elements of this integration, and are beginning to be manifested in terms of standards[103].

---

100 Without prejudice with regard to sites included within the limits of the state's territorial waters and to the powers of management and conservation of resources granted to the state by delegation; see article 20, regulation (EU) no. 1380/2013, cited above. On this point, see a circular dated April 30, 2013 from the minister of Ecology, Sustainable Development, and Energy relative to the inclusion of professional maritime fishery activities as part of the development, or revision where applicable, of documents laying out the objectives for Natura 2000 sites where these activities are carried out, NOR: DEVL1305078C.
101 Article L.921-10, 1st paragraph, CRPM.
102 For more information, we refer readers to Chapter 4 of [GRO 14] and Chapter 8 of [CUR 15].
103 See Directive 2014/89/EU of the European Parliament and of the Council of 23 july 2014 establishing a framework for maritime spatial planning, OJEU n° L 257/135 of august 28, 2014.

For its part, France is poised in 2015 to move toward finalizing the definition of a "National Maritime and Coastline Strategy[104]" aimed at achieving the "sustainable use of the seas and the conservation of marine ecosystems".

Thus, fishery finds itself part of a complex normative system whose objectives it must take on, yet without this system (for obvious reasons) being substituted for multiple aspects of the founding sectorial policy. In practice, fisheries policy must become the instrument of application for the objectives of the "strategy for the marine environment" directive in the fisheries sector. The state, by virtue of the powers it exercises in matters of management, will be bound to take all possible measures to minimize incidences of fishery on ecosystems while conferring fishing requirements on other coastal activities.

4.2.1.4. *Institutional framework*

Beyond the definition of the "maritime fishery" subject, it is important to emphasize the heterogeneity of the systems applicable in matters of resource management and conservation depending on the local authorities doing the planning, as well as the specific natures of the decision-making mechanisms instituted in order to regulate these matters.

4.2.1.4.1. Background information

The exercise of fishing activities in France is supervised by a set of texts, the rationalization of which, begun in 2010, has just been completed[105]. Codification was accomplished via the creation of a new book of the rural and maritime fisheries code (RMFC). The application of these regulations remains a fairly delicate matter[106] given the fact that it is still necessary to take into account the existence of special systems applicable to overseas collectivities.

First and foremost, we must delineate the field of application of book IX of the code entitled "Maritime fishery and marine aquaculture". The initial

---

104 "Cadre de référence pour la protection du milieu, la valorisation des ressources marines et la gestion intégrée et concertée des activités liées à la mer et au littoral", see article L.219-1 (and R.219-1-1) of the environmental code.
105 Decree no. 2014-1608 of December 26, 2014 relative to the codification of the regulatory section of book IX of the rural and maritime fisheries code, OJRF of December 27, 2014, p. 22407.
106 See *infra* section 4.2.2.

article refers to "maritime fishery; that is, the catch of animals and harvesting of marine plants, at sea and in saltwater part of rivers, streams, ponds, and canals[107]".

Maritime fishery is thus considered to constitute a whole, without distinction between professional and recreational activities, or between shipboard and on-foot fishing, though each of these activities is subject in practice to specific regulations. The saltiness of the water constitutes the determining criterion[108].

For streams and rivers flowing directly or indirectly into the sea, decrees set the points at which the water stops being salty, which constitute theoretical limits traced more or less arbitrarily that are used to delineate the domain of maritime fishery[109]. These limits do not make presumptions about the presence or absence on either side of these limits of fish species. Within these limits, the inland fisheries police have authority. With regard to saltwater ponds, these communicate in principle directly and naturally with the sea.

Book IX also applies to "the farming of animals and of marine plants", the various subsidiaries of which – fish farming, shellfish farming, shrimp farming and seaweed farming – are generally grouped under the generic term of marine aquaculture or "marine farming". Part of a global policy on maritime fishery and aquaculture, aquaculture is fully incorporated into all of the codified legislation but is nevertheless the subject of specific provisions, particularly having to do with the conservation and management of fish resources[110] and the professional body of shellfish farming[111].

4.2.1.4.2. Special systems

In France, the general scheme for the management and conservation of resources gives way, in overseas collectivities, to special systems

---

107 Article L.911-1 RMFC; formula taken from the abrogated legislative decree of January 9, 1852 on the exercise of maritime fishery.
108 See article D 911-2 RMFC.
109 See appendix 1 of book IX RMFC.
110 See section II, chapter 3 of book IX RMFC.
111 See section I, chapter 2, paragraph 2 of book IX RMFC.

determined on the basis of the provisions of the French constitution[112] and EU treaties[113].

Territorial authorities regulated by article 73 of the Constitution

This article stipulates that "in overseas departments and regions, laws and regulations are fully applicable under the law", though they may be subject to adaptations "taking into account the specific characteristics and limitations of these territorial authorities".

Thus, in the overseas departments and regions of Guadeloupe, Réunion, Guyana and Martinique[114], the administrative region has competence in matters of the management and conservation of marine biological resources "subject to France's international commitments, respect for community competence, and in compliance with the common fisheries policy[115]". This is also the case for the department of Mayotte, a unique collectivity exercising the competences allocated to overseas departments and regions.

Additionally, the regional council of each of these regions must be consulted for recommendations for all projected international agreements involving the exploration, exploitation, conservation or management of biological and non-biological natural resources in the exclusive economic zone (EEZ) of the French Republic in the open sea off the coasts bordering it[116].

Territorial authorities regulated by article 74 of the Constitution

These collectivities have a status which takes into account the interests proper to each of them within the French Republic. They are subject, depending on case, to the principle of legislative identity (Saint Barthelemy and Saint Martin) or the principle of legislative specialty (French Polynesia,

---

112 See article 72-3 of the Constitution.
113 See also the regulatory provisions applicable specifically to overseas territories: CRPM, regulatory section, section V as well as article R. 911-3 relative to state authorities competent to take measures of application in matters of maritime fishery in these various collectivities.
114 Since March 2014, the two latter collectivities have become uni-collectivities with competences granted to overseas departments and regions, and where reference to the department or region is replaced by reference to the territorial collectivity (see constitutive act no. 2011-883 of July 27, 2011 relative to collectivities regulated by article 73 of the Constitution and law no. 2011-884 of July 27, 2011 relative to the territorial collectivities of Guyana and Martinique, OJRF of July 28, 2011, p. 12818 and p. 12821, respectively).
115 See article L.551-1 of the rural and maritime fisheries code referring to article L.4433-15-1 of the general code of territorial collectivities (CGCT).
116 See article 4433-14 CGCT.

Wallis and Futuna, and St. Pierre and Miquelon), which means in this case that regulations applicable to the territorial collectivities of metropolitan France must be expressly applicable here as well. Thus:

– the collectivities of Saint Barthelemy and Saint Martin regulate and exercise "the right to explore and exploit biological and non-biological natural resources in inland waters, in particular safe harbors and ponds, in the sub-soil, and in waters above territorial waters and the economic exclusivity zone in compliance with France's international commitments and the competences of the state[117]";

– the competence of French Polynesia, though set by constitutive act[118], is comparable on all points to that of the collectivities above;

– on St. Pierre and Miquelon, "the state concedes to the territorial collectivity, under the conditions set by a set of specifications approved by decree by the Council of State accepted after recommendation by the territorial council, the exercise of competences in matters of the exploration and exploitation of biological and non-biological natural resources on the seafloor, its subsoil, and the overlying waters[119]";

– on the islands of Wallis and Futuna, the territorial assembly may undertake deliberations involving regulations in matters of "marine fishing, provided that this will not infringe upon the provisions of the code of rural and maritime fishery, on the system of territorial waters, or on general laws and regulations relative to high-seas fisheries[120]".

The specific case of the French Southern and Antarctic Territories and Clipperton

The territorial collectivity of the French Southern and Antarctic Territories (FSATs) – all of these lands form a single overseas territory – and Clipperton Island are subject, according to the last paragraph of article 72-3 of the French Constitution, to a specific legislative system and organization determined by the law:

---

117 Articles LO 6214-6 and LO 6314-6 CGCT, respectively.
118 See article 47 of constitutive law no. 2004-192 of February 27, 2004 relative to the autonomous status of French Polynesia (1), OJRF no.52 of March 2, 2004, p. 4183.
119 Article LO 6414-3 CGCT.
120 See article 40 of decree no. 57-811 of July 22, 1957 relative to the remits of the territorial assembly, the territorial council and the prefect of the islands of Wallis and Futuna, OJRF of July 23, 1957, p. 7252.

– The former are subject to the principle of legislative specialty. The exercise of fishing is regulated there by the dispositions of the rural and maritime fisheries code[121].

– Clipperton Island – which is not categorized as a collectivity but is also subject to the principle of legislative identity[122] – is under the direct authority of the government. The overseas minister in charge of administration delegates this task to the High Commissioner of the Republic in French Polynesia, as the government's representative.

New Caledonia

Governed by heading XIII of the French Constitution, this collectivity is subject to the principle of legislative specialty. By virtue of a constitutive law of March 19, 1999[123], New Caledonia is competent in matters of the "regulation and exercise of rights of exploration, exploitation, management, and conservation of biological and non-biological natural resources in the economic exclusivity zone".

For the sake of completeness, let us emphasize that heading V of book IX of the rural and maritime fishery code also specifies various legislative provisions concerning each of these collectivities.

The European Union and French overseas authorities

The application of EU law, and particularly of the provisions of the CFP, to these various authorities varies according to their status with regard to the treaties of the Union.

The provisions of the treaties of Union are applicable in ultraperipheral regions (Martinique, Guadeloupe, French Guyana, Réunion, Mayotte[124] and St Martin), where required, by means of specific measures developed by the European legislature in accordance with article 349 of the TFEU.

---

121 See CRPM article L.981-1 and f. and articles R 958-1 and f.

122 See article 9 of law no. 55-1052 of August 6, 1955 relative to the status of the French Southern and Antarctic Lands and of Clipperton Island, OJRF of August 9, 1955, p. 7979.

123 See article 22 of law no. 99-209 of March 19, 1999 relative to New Caledonia (1), OJRF no. 68 of March 21, 1999, p. 4197.

124 Mayotte acceded to the status of an ultraperipheral region of the EU on January 1, 2014 (see the European Council decision of July 11, 2012).

As for the overseas countries and territories targeted by appendix II of the TFEU (New Caledonia, French Polynesia, the French Southern and Antarctic Lands, the Wallis and Futuna Islands, Mayotte, and Saint Pierre-et-Miquelon), to the ranks of which St. Barthelemy was added on January 1, 2012, they are subject to a special system of association defined in Part Four of this treaty. The general provisions of the treaty, as well as of secondary law, are not applicable unless express reference is made[125]. In fact, the Union policy of resource management and conservation is not generally applicable in these places.

Finally, with regard to Clipperton Island, which is not mentioned in article 355 of the TFEU relative to the field of application of treaties, or by article 52 of the TEU, which provides for the application of treaties to the French Republic, the TFEU is not applicable[126].

4.2.1.4.3. The close collaboration between administrative authorities and professional bodies

Fisheries policy is implemented by the maritime fisheries and aquaculture directorate (MFAD)[127] under the authority of MEDDE[128] assisted by a consulting body, the Higher Advisory Council on fisheries, aquaculture and seafood-related policy[129], and a scientific and technical liaison committee for maritime fishery and aquaculture[130]. When decisions by public entities made in application of national or EU legislation relative to maritime fishery and marine aquaculture have an impact on the environment, they are subject to "public participation[131]". The minister also has special competences allowing him to make compulsory deliberations of the national committee on maritime fisheries and maritime farming relative to measures for the conservation and management of fisheries resources[132].

---

125 See CJCE of February 12, 1992, Leplat, aff. C-260/90, point 10, Rec. p. I-643.
126 See in this context the response of Mr. Gaston Thorn, president of the Commission, to written question no. 1007/84, JOCE no. C.62, March 11, 1985, p. 34.
127 DPMA in French "Direction des Pêches maritimes et de l'Aquaculture ".
128 MEDDE in French "Ministère de l'Ecologie, du Développement Durable et de l'Energie".
129 Article L.914-1 RMFC.
130 Article L.914-2 RMFC.
131 Article L.914-3 RMFC, under the conditions specified by the environmental code.
132 See in particular articles R.912-14 and R.912-15 RMFC.

In each of the four major maritime regions defined by decree[133], a marine environment inter-regional director exercises, under the authority of competent regional prefects, authority relative to the regulation of the exercise of fishing activities (professional or recreational) and to the control of the activities and management of regional maritime fishing committees and regional shellfish-farming committees[134].

Some regional prefects[135] exercise general competence in matter of the conservation and management of fishing resources. They may take regulation in this field and hand down decisions making compulsory the deliberations of regional committee councils.

Consequently, it appears that professional organizations, though private in terms of regulatory powers, participate in a decisive manner in the definition and implementation of resource conservation and management policies[136]. It falls to committees for the professional organization of fishery and marine farming to "participate in the development of regulations in matters of management of fish resources and the harvest of marine plants", as it does to committees for the professional organization of shellfish-farming to harmonize both methods of production and exploitation in the field of shellfish-farming and good cultural practices. Standards are developed as part of a co-management process between professional organizations and competent administrative authorities, which will be described later.

For their part, "producers' organizations" originating from European legislation[137] ensure a better use of the subquotas assigned to them[138] and participate in regulatory activity as well, through the issuance, under the supervision of administrative authorities, of fishing authorizations aimed at

---

133 Decree no. 2010-130 of February 11, 2010 relative to the organization and missions of inter-regional marine directorates (article 3), OJRF no. 36 of February 12, 2010, p. 2507.
134 Articles L.912-1 and f. and L.912-6 and f. RMFC.
135 Articles R.911-3 and R.911-4. There are six of these in metropolitan France and five in the overseas departments; for these overseas departments, particularly Mayotte, see also article L.4433-15-1 CGCT. In the other overseas collectivities, see the especially designated authorities, article R.911-4, point II.
136 Articles L.912-2 and L.912-7 RMFC.
137 Regulation (EU) n° 1379/2013, above-mentioned.
138 Article L.921-5 RMFC.

species subject to an authorized catch total or to catch quotas in application of European regulations[139].

In considering professional organizations, we must not omit "prud'homies", ancestral bodies arising from Mediterranean fishing tradition, which are regulated by a decree dated November 19, 1859. Among other powers, these bodies are authorized to take measures in order to avoid usage conflicts between professions[140].

The departmental prefect, finally, exercises general competence of common law in the field of marine farming, but has only residual regulatory powers in matters of fishery[141]. The implementation of marine and coastal policy, particularly with regard to fishing activities and marine farming, is overseen, under its authority, by the departmental territory and marine directorate[142].

4.2.2. Instruments of French fishery policy

In compliance with international law, the state exercises full competence to manage fishery activities exercised in the waters surrounding various overseas territorial collectivities. However, for the most part, French conservation of fish resources is framed by the regulations of the EU, which has exclusive competence in this domain, and in principle grants the state only delegation or execution powers. It is important from this point of view to delineate the field of intervention for national authorities before examining the system of "fishing rights" that is at the core of the conservation policy.

4.2.2.1. Area of state intervention

The state holds power of conservation and management within the 12 nautical mile zone situated beyond its baselines and within which the EU also allows it to grant privilege of access to its nationals. This 12 mile limit constitutes a line of sharing that is quite practical, in that it distinguishes within the Union's fishing zone a space within which each state enjoys

---

139 Article L.921-2 RMFC.
140 For more information on prud'homies, we refer readers to Chapter 1 of [FAG 14].
141 In particular, article D.922-22 CRPM.
142 Article 3, IV of decree no. 2009-1484 of December 3, 2009 relative to interministerial departmental directorates, OJRF of December 4, 2009, text no. 1.

specific prerogatives. It must not, however, cause errors that would result in this area being exempt from EU competence, since the general principles and regulations decreed by the Union are fully applicable in these areas, and it can impose specific regulations in them that it judges appropriate. Thus, measures to limit fishing possibilities (in terms of both catches and efforts) have a general scope that exceeds these spatial considerations.

Moreover, French legislation does not refer to the 12 mile zone[143]. Rather, it establishes a clear distinction between management measures according to whether they are aimed or not at species subject to "a total allowable catch or to catch quotas in application of European regulations[144]". The second hypothesis covers the field of measures arising effectively from the powers exercised by national authorities, and with regard to which the law remains relatively evasive[145]. Considering that in practice, most of these measures are adopted by virtue of procedures combining administrative authorities and professional committees, articles R.912-14, R.912-15, R.912-31 and R.912-32 of the rural and maritime fisheries code give a fairly clear view of their impacts.

Moreover, if fishing activities are exercised in accordance with Union regulations, the implementation of legislation requires – or enables – the state to act. In reality, the exclusive competence exercised by the Union in matters of conservation does not require it to regulate every aspect of this conservation. Thus, activities such as underwater fishing, the exploitation of algaes, fishing from shore and recreational fishing as well as the organization of activities in the space are – yet again – exempt from the normative powers of the Union, which entrusts the state with the power to regulate them[146].

State action[147] carried out at the national or regional level by administrative authority alone[148] or, most often, in partnership with professional

---

143 Unlike regulation that specifically sets out measures for inland waters and territorial waters (see in particular articles D.922-18 and D.922-28 CRPM), and even the "three-mile coastline" (article D.922-17). These provisions implement powers delegated by the Union to member states in the context of article 20 of regulation no. 1380/2013, cited above.
144 See articles L.921-2 and L.921-2-2 RMFC.
145 See article L.921-2-2, 2nd paragraph RMFC.
146 See RMFC, regulatory section, title II, chapter 1, section V, "Régimes particuliers d'autorisations de pêche".
147 *Supra*, section 4.1.
148 Generally after consulting with the professional organization (see articles L.921-2-1 and L.921-2-2 RMFC and articles 3 and 14 of decree no. 2011-776).

organizations as part of a "co-management" system, consists principally of the institution and implementation of a system of fishery rights.

4.2.2.2. System of fishing rights

This concerns control of both catches and fishing efforts.

4.2.2.2.1. Control of catches

This is based on a quota system and tends to limit quantities caught.

Most major commercial species are managed on the basis of total admissible catches (TACs) set by the Council of the EU and subsequently distributed among member states in the form of quotas[149]. Competent regional prefects have equivalent powers with regard to species not covered by European regulations and fished in inland waters and in French territorial waters[150].

The minister then proceeds with the division of quotas into subquotas allocated either to POs – or their unions – which then ensure their management[151], or to vessels or groups of vessels when the latter do not belong to a PO[152].

Finally, the administrative authority – minister or competent regional prefect – can make compulsory the deliberations of professional committees relative to the limitation of the catch volumes of certain species[153].

4.2.2.2.2. Control of fishing efforts

This relies on two types of instruments: quotas and prior authorizations.

---

149 See *supra*.
150 See article R 921-37 RMFC.
151 Modes of managing subquotas by POs vary, ranging from simple collective management (after distribution of subquotas between members, with the PO simply monitoring the consumption of resources) to individual management (the PO distributes the available resource among its members and per vessel on the basis of specific criteria). Between these two extremes, there are a variety of hybrid solutions through which the PO applies individual limitations – often applicable per group of vessel – on the basis of variable criteria as part of collective management. These modes of functioning should be re-evaluated given the provisions set by the new CFP regulations.
152 See article L.921-4 RMFC.
153 Voir *supra*.

Fishing effort quotas

This method of control is the global result of Union legislation requiring member states to implement "measures to adapt the fishing capacity of their fleets in order to achieve a stable and sustainable balance between fishing capacity and their fishing possibilities[154]". A plan for adapting the fishing capacity of the fleet to fishing opportunities is generated annually[155]. Limiting fishing effort has resulted in France in intermittent ordinances distributing the efforts allocated by the Union to the state among certain fisheries[156] and creating a national fishing effort system for other fisheries[157].

The allocation of fishing efforts among recipients is done by competent administrative authorities using identical methods to those used for quotas[158].

Finally, the competent administrative authority – minister or regional prefect – can also make compulsory the deliberations of professional organization committees[159] relative to the balancing of fishery capacities with available resources, particularly by means of an adjustment of fishing efforts[160].

Prior authorizations

Control of access to fisheries resources is also established – often in a cumulative manner – via the institution of a system of prior authorization.

---

154 See article 22 of (EU) regulation no. 1380/2013, cited above.
155 Article R.921-9 RMFC.
156 For example, a decree of February 12, 2014 relative to the distribution of fishing effort quotas allocated to France as part of the reconstitution of certain deep-water and cod stocks in zones ICES III a, IV, VI a, VII a and VII d as well as in community waters of zones ICES II a and V b and as part of the exemption specified in cod stock reconstitution zones for the year 2014, OJRF of February 15, 2014, p. 2675.
157 For example, a decree of January 28, 2013 involving the creation of a fishing effort system for professional fishery using trawling nets in the Mediterranean Sea by vessles flying French flags, OJRF of February 27, 2013, p. 3275.
158 Article L.921-4 RMFC.
159 The committees are the manifestations of the professional fishing organization, whose mission consists in particular of the organization of a balanced management of resources
160 Articles R.912-14, R.912-15, R.912-31 andR.912-32 RMFC, cited above.

Broadly speaking, there are two types of authorizations: the first authorization has to do with the use of vessels flying their flags for fishing, while the second authorization concerns the exercise of specific fishing activities.

Authorizations relative to the use of vessels flying their flags for fishing

A fishing vessel license (PME)[161], as its name suggests, is a prior authorization required by French legislation[162] for all professional fishing vessels before they are declared fit for operation. Attached to the vessel, it is issued by the regional prefect of the ship's registration location, after consultation with a regional commission for maritime fishery and marine aquaculture. The minister responsible for fisheries, taking into account the plan of adaptation of the fishing capacities of the fleet to fishing opportunities, and the evolution of the fishing fleet noted throughout the previous year, sets the available contingent each year (expressed in engine power and tonnage) on the basis of guidelines established by European regulations.

The obtaining of this fishing vessel license is a precondition for the issuance of a "fishing license", an authorization required by European legislation. This license concerns any Union fishing vessel intended for the commercial exploitation of living aquatic resources and becomes part of the file on the community fishing fleet[163]. Both types of authorizations are issued and managed by the state.

Authorizations relative to the exercise of fishing activities

French law stipulates that all professional fishing activities may be subject to authorization, which is generally the case. Authorizations "are intended to enable an individual or legal entity, for a determined vessel, to exercise these activities during periods, in zones, for species or groups of species, and, where applicable, with determined fishing gears and for determined amounts of catches"[164]. These authorizations are provisional, covering a maximum period of 12 months, and are non-transferrable[165].

---

161 PME or Permis de mise en exploitation in French.
162 Articles L.921-7 and R.921-7 and f. RMFC.
163 Regulation (EC) n° 1224/2009, above-mentioned.
164 Article L.921-1 RMFC.
165 Consistent with the collective heritage status of halieutic resources.

However, unlike the quota system, the authorization system does not necessarily lead to a limitation of fishing efforts unless accompanied by a quota system.

The definition of the code of regulations may cover several types of authorizations. According to the rural and maritime fisheries code[166], "fishing authorizations for species subject to a total admissible catch (TAC) or to catch quotas in application of European regulations are issued by the administrative authority or, under its supervision, by POs or their unions. For other species, fishing authorizations are issued by the administrative authority or, under its supervision, by the national or by regional committees of maritime fisheries and marine farming". In practice, with regard to authorizations managed by administration, these are issued, after consultation with the national committee, by the MFAD or the competent regional prefect – more commonly, the marine environment inter-regional director upon delegation of the latter. With regard to authorizations managed by POs, fishing authorization is issued by the organization to a shipowner for a given vessel and belonging[167].

Thus, the law distinguishes authorization systems according to whether they concern managed or non-managed species in the context of a TAC (or quota) system of European fishery.

Authorizations for fish species not concerned by the European TAC or quotas system

Vessels must hold European fishing authorizations with regard to fisheries – or fishing zones where these activities are authorized – fulfilling certain criteria[168]. In principle, vessels less than 10 meters long are exempt, though the state can revoke this exemption. Moreover, there is nothing prohibiting national authorities from creating a system of specific

---

166 Article L.921-2 RMFC.
167 For example, a decree of December 28, 2012 relative to the creation of European fishery authorizations for certain fisheries not subject to quotas but subject to a multi-year management plan adopted by the European Union, OJRF of December 30, 2012, p. 21216, spec. article 2.
168 Regulation (EC) n° 1224/2009, spec. article 7, above-mentioned.

authorizations for species under European TAC system but not falling within the context of the previous criteria[169].

Authorizations for fish species not concerned by the European TACs or quota systems

The administrative authority may subject these activities to authorization with regard to vessels flying the French flag[170]. These are generally regional licenses[171] that are most often linked to specific instruments for the management of a resource or a specific "métier[172]" and which, consequently, may involve various specific measures such as fishing period, fishing zone(s), vessel characteristics (limitations of size or engine power), type of fishing gear (and its characteristics), catch volume, etc.

These licenses are issued by the administrative authority or, under its supervision, by the *ad hoc* committees of the professional organization according to the co-management mechanism previously described. The use of this second solution is generally favored by the administrative authority. Rather than the national committee, it is regional committees that use this prerogative. Thus, the only regional maritime fishery and marine farming committee in Brittany undertakes several dozen deliberations each year, generating fishing licenses for nearly 30 species or "activities" practised under its territorial responsibility.

While these licenses generated by the national committee may cover the totality of the waters under French sovereignty or jurisdiction, those issued by regional committees concern only the activities practiced in inland waters and territorial waters falling under its geographical responsibility.

---

169 For example, a decree of December 9, 2011 supervising the fishery of langoustines (*Nephrops norvegicus*) in zone ICES VIII a, b, d and e, OJRF of December 23, 2011, p. 22025.

170 And under certain conditions, vessels flying a foreign flag; see article L.921-9 CRPM.

171 Distinct from the "licenses" of European legislation.

172 Métier: "a group of fishing operations targeting a similar species using similar gear, during the same period of the year and/or within the same area and which are characterized by a similar exploitation pattern" EUR-Lex Europa.

Fishing authorization quota system

It now becomes necessary to distinguish authorization systems according to whether or not they implement a *numerus clausus*[173].

For non-quota fishing activities, any vessel may file a request for authorization, subject, if applicable, to compliance with the conditions of conferral of the authorization or fishing possibilities – such as the availability of a quota or a subquota for catches and/or effort – provided for by regulations.

For fishery activities constrained by a fishing authorization quota system, only vessel-vesselownership couples fulfilling the necessary conditions specified by regulations may request authorization. Thus, there is a list of vessel-ownership couples "eligible" for quota-including authorizations issued by the minister in charge of fishery[174] – via the MFAD – for each of the authorization systems in force. An "ineligible" ship-owner couple can, however, obtain an authorization as part of a transfer request approved by the administrative authority.

4.2.2.2.3. Criteria for the conferral of fishing rights

Though the law[175] does not specify *criteria* having to do with the details of the distribution of fishing quotas, it is advisable to refer to the regulatory provisions of the code (RFMC) that provide vital clarifications on this subject[176]. In the absence of specific indications, the definition of these *criteria* as they emerge from these provisions is also valid for the conferral of prior authorizations[177]. These are, in brief:

– "anteriority right of producers[178]," which play a leading role in the conferral of rights, are "historical references" having to do with fishing activities and established on the basis of data declared by the fishing masters. They constitute, according to the text, a "method of calculation" used to

173 See an MFAD circular dated December 24, 2012 on the details of management relative to European and national fishery authorizations for the management year 2013, not appearing in the OJRF.
174 See articles R.921-20 and f. RFMC.
175 See in particular article L.921-4 RFMC.
176 Articles R.921-35 and f. RFMC.
177 See article L.921-2 RFMC, paragraph 1 and article R.921-21, paragraph 1, CRPM.
178 Articles R.921-38 and R.921-39 RFMC. The "producer" is the owner of a professional fishing vessel flying the French flag, registered in the European Union and declared active in the files of the Union fishing fleet.

proceed with the distribution of quotas "and not a right allowing a claim of these quotas". This latter point is obviously crucial, though it is subject to caution;

– "orientations of seafood market[179]" which enable the minister in charge of fisheries to regulate the contributions of producers in order to add value to landings; for example, by setting periodic catch limits and/or landing limits or by temporarily suspending the fishing of a specie subject to a quota. These measures may be decided depending on the circumstances of various "professions" and fishing equipment, registration sites of producers' fishing vessels, fishery zones or landing sites;

– "socio-economic balances[180]", finally, which enable the minister in charge of maritime fishery to redistribute a quota by imposing additional *criteria* for access to fishery, where applicable by means of a system of specific fishing authorizations, taking into account in particular "professions" and fishing equipment, registration sites of producers' fishing vessels; fishing zones and landing sites. This minister may also decide to put producers' anteriority rights "in reserve", temporarily and under some circumstances, by placing them into a "reserve of anteriority rights and quotas" in order to reconfer them later depending on various objectives, in order particularly to foster an influx of new entrants to the profession.

The Council of State has clarified that the minister in charge of fishery, when he/she decides to redistribute fishing quotas, must combine the three criteria[181]. However, as regulatory provisions do not impose any particular guidelines for taking into account of these various *criteria*, he/she consequently possesses discretionary competence in this area.

To conclude, we will note that while control of access to resources constitutes a decisive element of the national fisheries resource conservation policy, it does not act alone. In reality, the EU does not prohibit any of its member states from adopting general measures for the conservation of stocks in waters under its sovereignty or jurisdiction, subject to the condition that these are applicable only to its own vessels – or, barring this, to individuals resident in that state – and that they are not less strict than those imposed by the Union. Thus, states have a relatively large amount of

---

179 Article R.921-49 RFMC.
180 Article R.921-50 RFMC.
181 Notably the Council of State, meeting of 3rd and 8th subsections, 7/19/2011, no. 329141, mentioned in the Lebon tables.

flexibility in their territorial waters[182] and in matters concerning activities that do not fall within the field of application of Union regulations[183]. They may take "technical measures" (minimum reference size, prohibited fishing methods, characteristics of fishing gears, etc.); as well as order and precaution measures[184] (intended to regulate the exercise of fishing activities and to foster compatibility between "métiers") and they may also set temporal and spatial restrictions (the prohibition or regulation of fishing of certain species or with certain gears in certain areas; the classification of natural shellfish deposits, measures for the delineation of reserves or areas in which all fishing is prohibited, etc.)[185] – none of these, however, are measures that call for particular commentary.

4.3. Bibliography

[CLE 08] CLEACH M.P., L'apport de la recherche à l'évaluation des ressources halieutiques et à la gestion des pêches, rapport, Office parlementaire d'évaluation des choix scientifiques et technologiques, no. 1322 de l'Assemblée nationale, December 2008.

[CUR 15] CURY Ph., BERTRAND A., BERTRAND S. *et al.*, "L'approche écosystémique des pêches: réconcilier conservation et exploitation", in MONACO A., PROUZET P. (eds), *Diversité et fonctions de systèmes écologiques marins*, Chapter 8, ISTE Editions, London, pp. 343–426, 2015.

[FAG 14] FAGET D., SACCHI J., "Fishing in the Mediterranean, past and present: history and technical changes", in MONACO A., PROUZET P. (eds), *Development of Marine Resources*, Chapter 1, I ISTE, London and John Wiley & Sons, New York, pp. 1–55, 2014.

[GRO 14] GROS P., PROUZET P., "Impact of global changes on the dynamics of marine living resources ", in MONACO A., PROUZET P. (eds), *Ecosystem Sustainaility and Global Change*, Chapter 4, ISTE, London and John Wiley & Sons, New York, pp. 113–212, 2014.

[GUE 11] GUÉDON L., Vouloir une politique de la pêche pour la France, Report to the Prime Minister, March 2011.

---

182 Regulation no. 1380/2013, article 20, above mentioned.

183 See in particular "Régimes particuliers d'autorisation de pêche", RFMC, regulatory section, part II, chapter 1, section V.

184 Articles L.921-2-1 and L.922-2, articles R.922-24 and f. RFMC.

185 Article L.922-2 and R. 922-8 RFMC.

[HAR 68] HARDIN T., "The tragedy of commons", *Science*, vol. 162, no. 3859, pp. 1243–1248, 1968.

[LEB 11] LE BIHAN C.D., CUDENNEC A., "La politique commune de la pêche", in BLUMANN C. (ed.), *Politique agricole commune et Politique commune de la pêche*, Commentaire Mégret, Editions de l'Université de Bruxelles, Brussels, 2011.

[MON 14a] MONACO A., PROUZET P., *Development of Marine Resources*, ISTE, London and John Wiley & Sons, New York, 2014.

[MON 14b] MONACO A., PROUZET P., *Ecosystem Sustainaility and Global Change*, ISTE, London and John Wiley & Sons, New York, 2014.

[MON 15] MONACO A., PROUZET P., *Marine Ecosystems*, ISTE, London, 2015.

[PRA 96] PRAT J.L., "Le droit communautaire et le droit français des pêches maritimes en quête de quotas individuels transférables", in FALQUE M., LAMOTTE H. (eds), *Droits de propriété, économie et environnement. Les ressources marines*, Dalloz, Paris, pp. 88–96, 1996.

Marine Renewable Energies: Main Legal Issues

5.1. Introduction

The renewed use of the sea, linked in particular with the boost in ocean renewable energy, calls for reflection on the development of the legal framework in which it operates. These new uses, located in marine areas, pose several legal questions which concern not only the recurrent themes of conflicts of use and sustainable development, but also the underestimated topics of maritime urbanization and the economic exploitation of common resources[1].

The different marine technologies for energy production (fixed or floating wind-turbines, wave or tidal generators, sea thermal power plants and osmotic energy) are not all at the same stage of maturity (for more details, see [PAI 14]). While some are already market-ready, others will be in the medium or more long term. Ocean renewable energies produce more power than and are less intermittent than their terrestrial counterparts, but are subject to multiple technical constraints. The construction of these installations is more complicated, long and expensive than on land. Thus, for a fixed wind-turbines, the construction schedules are in the range of 2–4

Chapter written by Nicolas BOILLET and Gaëlle GUEGUEN-HALLOUET.

1 Currents, air, waves, sea thermal energy, marine biomass, tides and salinity gradient energy. See the inventory of these sources in the "Report of the Secretary-General of the United-Nations on ocean renewable energies" 4th April 2012, ref A/67/79.

years at sea, as opposed to a few months on land – the anchoring work being very significant and the installations themselves larger in size. The amounts invested and the risks inherent to offshore wind-turbine projects thus require a secure legal framework.

The development of renewable marine energies has its origin in the United Nations' Framework Convention on Climate Change[2], adopted on 9 May 1992, and in the additional protocol, adopted in 1997 in Kyoto. These texts outline the objective to stabilize the emissions of greenhouse gases into the atmosphere. Simultaneously, diverse measures adopted by the European Union [GRA 11] have paved the way for renewable energies. Directive 2001/77/EC of the European Parliament and the Council of 27 September 2001 on the promotion of electricity produced from renewable energy sources in the domestic electricity market[3] outlines most of the necessary measures. The directive, adopted in the wake of the "White Book" that set out a european community strategy and plan for renewable energies[4], recommends the increase in the production of renewable energies.

Although a number of member states (such as Spain, Sweden, the Netherlands, the United Kingdom[5], Denmark[6] and Norway[7]) quickly set out to execute their European obligations for the development of renewable energies on land as well as on sea, France has lagged behind. Let us note that pursuant to the Grenelle I[8], France has set itself the goal of having 23% of the total consumed energy produced from renewable sources by 2020.

In order to achieve this goal, the first action to be taken was to install marine wind farms of 1,000 megawatts (MW) on the 31st December 2012. Further farms of 6,000 MW will be added, for a total production of wind

---

2 JOCE L 33, 7 February 1994.

3 JOCE L 283, 27 October 2001.

4 Energy for the Future, Renewable Sources of Energy, COM(97)599 final, 26 November 1997.

5 The UK set itself a goal of 15% of renewable energy in the final energy consumption and 30% of electricity produced by 2020, among which 1% would be offshore tidal and 44% offshore wind power.

6 Wind supplying 23% of electricity

7 Member state of the European Economic Community.

8 Law 2009-967 of 3 August 2009 on the timetable for implementation of the Grenelle Round Table on the Environment, JORF no. 179, 5 August 2009, p. 13031.

energy of 25,000 MW by 31st December 2020[9]. According to the Union of Renewable Energies, this will mean the construction of 100–120 wind-turbines each year for 10 years, representing more than 1.5 billion euros of investment each year. France is currently ranked third in Europe for installed generating capacity from on-shore wind-turbines. However, even though the surface area of France's territorial waters and exclusive economic zone is greater than 10 million km^2, nearly 20 times the surface area of the national territory, no offshore capacity is currently being exploited. Nonetheless, the construction of offshore wind-farms opens new opportunities for the development of renewable energy, each turbine capable of a maximum output of 5 MW compared to 3 MW for on-shore turbines.

To this end, the Parliamentary Commission of common information on wind energy[10] has deemed it essential "to assert France's ambition in this field and to support the orientations of the Interministerial Committee for the Sea, as defined the 8 December 2009, for the emergence, on a national scale, of a industry-leading, exporting scientific infrastructure". More recently, the Ministers for Ecology and Industrial Recovery commissioned a study on ocean renewable energies in order to create an action plan for the achievement of the national objectives of diversifying renewable energies at sea. Among the recommendations in this report, published on 16 March 2013 [BOY 13], it is interesting to note that its authors found that Marine Renewable Energies (MRE) projects are subject to a complex range of legal rules, a likely source of delays, if not litigation. The authors state that "this legal framework should be simplified whilst conserving a high level of environmental protection".

Although, for many years, the French legal system did not consider the specificities of ocean renewable energies, the French government now seems to have become aware of the importance of establishing an adapted legal framework. Certain trends should push forward its development in the next few years. The "Blue Book" that resulted from the Grenelle round table meetings announced the objective "to clarify the regulations applicable to ocean renewable energies and to integrate them into the various decision-making levels[11]". The law no. 2010-788, adopted on 12 July 2010 [JOR 10], has

---

9 Decree of 15 December 2009 on the multiannual programming of investment of electricity production, JORF no. 8, 10 January 2010, p. 526.

10 Information report filed by the commission of common information on wind energy on the 31 March 2010.

11 Blue Book of commitments of the Sea Grenelle, 10 and 15 July 2009, Ministry of Ecology, Sustainable Development and Energy.

already provided a set of incentive measures for the development of offshore wind-turbines.

The boom of marine renewable energies is thus the result of proactive policies from a French nation that seeks to meet its international and European commitments (section 5.2). It entails the adoption of an adapted legal framework.

5.2. French policy for the development of marine renewable energies: foundations and instruments

The policy for the development of marine renewable energies is the result of the international and European commitments that France has signed up to (section 5.2.1). It makes use of legal instruments that ensure programming and planning (section 5.2.2).

5.2.1. The international and European foundations for the development of renewable energies

The UN committed very early to the development of programs for renewable energies and continues now in this direction[12]. However, it was only from 1992, and the Rio Earth Summit, that the first international conventions were adopted to target and oversee the development of renewable energies. These international texts, based on both the climate and the market (section 5.2.1.1), have more recently been translated into European law which gives them a binding power.

5.2.1.1. International texts based on the climate and the market

Before 1992, energy was, at the most, an indirect object of international law. The conventional legal texts were rare and the non-conventional texts were limited to a few case-law principles which were not specific to

---

12 It has committed to the promotion of renewable energies, notably following the conference of Rome in 1961. We should also note the initiative launched by the Secretary General Ban Ki-moon, on 7 November 2011, primarily supporting three interdependent objectives: to reach universal access to modern energy utilities, to double the rate of improvement of energy efficiency and to double the share of renewable energies in the world's energy mix. We should also note the Informal Consultative Process (ICP) on oceans and the law of the sea which had its 13th meeting in May 2012 and whose theme was ocean renewable energies (see [BOI 13]).

renewable energy [BOI 13a]. The motivation to develop renewable energies did not really appear until the moment when the issue of climate change became of international concern.

Until the Rio Earth Summit in 1992, energy did not constitute, in international law, a coherent object with its own legal regime. Carefully sidestepped in conventions on free-exchange for politico-strategic reasons, energy is considered in certain instruments of international environmental law. Although the sources of energy are not specifically mentioned, the United Nations' Stockholm Declaration on the Environment, of 16 June 1972, affirms the conviction of the signatories that "the capacity of the earth to produce vital renewable resources must be preserved and, wherever possible, restored or improved" (principle 2). Other international conventions adopted with a concern for the protection of the environment also have an impact on the development of ocean renewable energies. These essentially concern international texts on the protection of spaces and species[13] that have been directly transposed into national law or via European law. One example is the Paris Convention, of 22 September 1992, for the protection of the marine environment of the Northeast Atlantic. This text enshrines the application of the principle of precaution to artificial structures[14].

Before 1992, the United Nations convention on the law of the sea, signed on 10 December 1982 at Montego Bay, is the only international text that explicitly mentions marine renewable energies. Entering into force in French

---

13 This applies, for example, to the Ramsar Convention of 2 February 1971 on internationally important wetlands (such as water birds' habitats) and the Bonn Convention of 23 June 1979 on the conservation of migratory species of wild animals (according to article 4, requiring parties to endeavor to prevent remove, compensate for or minimize the adverse effects of activities that seriously impede migration). Similarly, the Bern Convention of 19 September 1979, adopted within the framework of the Council of Europe and entering into force 6 June 1992, aims to ensure the conservation of Europe's wildlife and the natural environment by interstate cooperation. It was ratified by the European Community by the Council Decision no. 82/72/EEC of 3 December 1982.
14 According to article 2 §2 of this convention: "2. The Contracting Parties shall apply: (a) the precautionary principle, by virtue of which preventive measures are to be taken when there are reasonable grounds for concern that substances or energy introduced, directly or indirectly, into the marine environment may bring about hazards to human health, harm living resources and marine ecosystems, damage amenities or interfere with other legitimate uses of the sea, even when there is no conclusive evidence of a causal relationship between the inputs and the effects".

law following the publication of decree no. 96-774, of 30 August 1996, this convention rules, for and between the 117 states, on "everything that concerns marine areas, from the question of delimitation to the control of the environment and the exploitation of the deep sea, including scientific research, commercial activities, technology and the settlement of disputes related to problems of the sea".

The Montego Bay Convention, therefore, recognizes coastal states' full sovereignty over their inland waters and territorial sea, subject to respect of the right of innocent passage, and of conventions that limit, one way or another, certain freedoms of action (articles 3 to 15).

It also recognizes the states' sovereign rights "of exploration and exploitation for economic purposes, in the exclusive economic zone, such as the production of energy from water, currents and winds" (article 56 §1). Furthermore, articles 77 to 80 contain similar provisions to the above, related to installations on the continental shelf. Article 87 relates to installations on the high sea and establishes coastal states' freedom (or not) to build artificial islands or other installations authorized by international law, subject to Part IV (continental shelf). Finally, the convention specifies, in article 206, that the development of ocean renewable energies should be implemented, while respecting the protection of the marine environment.

It should be noted that, before 1992, the number of non-conventional texts[15] was also small. Nevertheless, certain principles could be applied to renewable energies, even if they mainly concerned environmental protection. This is true for the customary principle of the permanent sovereignty over natural resources applicable to all, asserted from 1962 and restated by the Rio Declaration on the environment and development (article 2) [BOI 13].

Although it only mentions one source of energy – biomass – the United Nations Framework Convention on Climate Change [JOC 94] (UNFCCC), adopted in 1992, for the first time calls on its signatories, including the states of the European Union, to adopt national programmes for reducing greenhouse gas emissions, urging them to stabilize, by 2000, their emissions to the level of 1990. Following this, the Kyoto Protocol, adopted on 11 December 1997 [JOC 02], defines the means for tackling climate change. It

---

15 According to article 38 of the Statute of the ICJ, these involve customs, general principles of international law, case law and doctrine.

gives the objective of an overall reduction of greenhouse gas emissions, compared to 1990, of 5.2% between 2008 and 2012. The European Union committed to a reduction of 8%. Furthermore, the Kyoto Protocol points toward the development of renewable energies, that it considers as a means of achieving its goals.

During the same period, other international economic instruments have concerned renewable energies, although undoubtedly quite cautiously. For example, the Energy Charter Treaty adopted on 17 December 1994 by 49 states and entered into force on 16 April 1998[16]. This text, undoubtedly economic and financial in its scope, aims to establish "a legal framework in order to promote long-term cooperation in the energy field" and to "catalyze economic growth by means of measures to liberalize investment and trade in energy" (articles 1 and 2). Nevertheless, some of the provisions might, to quote Professor Boiteau, "be used as anchoring points for renewable energies".

Still in the economic sphere, the fundamental rules of the World Trade Organization relate to the trade of goods and services in the energy sector, including renewable energies[17]. The Doha round of negotiations was the first multilateral negotiations on trade and environment to include the energy sector, even if there is still a long way to go. These negotiations show the new willingness to adapt the multilateral trading system to the requirements of effectively tackling climate change, in particular by promoting exchanges on the subject of renewable energies.

5.2.1.2. A binding European commitment

Even though energy is a primary concern for European institutions[18], no provision in the treaty founding the European Economic Community made reference to it. Nevertheless, this has not prevented European judges from

---

16 Report on the Energy Treaty Charter, *Enerpresse*, p. 2, 22 January 2001.

17 A good example of this application is the current dispute between the European Union and Canada over the "Feed-in Tariff program". The European Union has demanded that consultations be opened with Canada over measures that affect the sale, purchase, transport, distribution or use in the national market of equipment for renewable energy production plants which prejudice imported equipment in favor of similar products from Ontari: Aff. DS 426.

18 European integration was originally founded on the Treaty of Paris of 17 April 1951 establishing the European Community of coal and steel, and on the treaty of 25 March 1957 establishing the European Community of nuclear energy.

qualifying energy as a commodity [CJE 64, CJE 94], which must benefit from free circulation in the common market. It may be concluded from this that the production of electricity is a commercial service.

As a result, certain measures were adopted in this field which have held until the institutionalization of an energy policy in the Treaty of Lisbon which entered into force on 1 December 2009[19]. These measure were either based: on the old article 100 EC[20] on the approximation of legislations, or on the old article 235 EC[21] (now classed as a flexibility clause, it permits intervention, in the absence of a specific legal basis, when the functioning of the common market is concerned), or finally, from 1986 onward, on the basis of the provisions on the environment[22].

All these measures formed the basis of the energy policy that is enshrined in the Treaty of Lisbon. They are the result of, on the one hand, the European market's progressive construction of a liberalized economic interest group, which aims to balance the opening up to competition and the public service missions[23]; on the other hand, they result from the adoption of European instruments that define a European strategy on renewable energies, and set production objectives in this field[24]. Here, it essentially involves

---

19 Article 194 of the TFEU (Treaty on the functioning of the European Union).

20 Now article 115 of the TFEU.

21 Now article 352 of the TFEU.

22 Now 191 to 193 of the TFEU.

23 This is the result of the adoption of three legislative packages in 1996, 2003 and 2009. Among the adopted texts, there is the directive no. 96/92/EC of 19 December 1996 concerning common rules in the internal market in electricity (JOCE no. L 27/2 of 30 Jan. 1997); the directive no. 2003/54/EC of 26 June 2003 concerning the common rules in the internal market in electricity (OJEU no. L 176/37 of 15 July 2003), repealing the directive 96/92/EC and regulation (EC) no. 1228/2003/EC, 26 June 2003 on conditions for access to the network for cross-border exchanges in electricity (OJEU no. L 176/1 of 15 July 2003); the directive no. 2009/72/EC of 13 July 2009 concerning the common rules in the internal markets in electricity (OJEU no. L 211/55 of 14 August 2009, p. 55), repealing directive 2003/54/EC – EP and EU Council, and regulation (EC) no. 714/2009, 13 July 2009, on access to the network for cross-border exchanges in electricity, repealing the regulation (EC) 1228/2003 and the regulation (EC) no. 713/2009 of 13 July 2009 establishing an Agency for the cooperation of energy regulators (OJEU no. L 211/15 of 14 August 2009).

24 Directive no. 2009/28/EC of 23 April 2009 on the promotion of the use of energy from renewable sources, modifying then repealing the directives 2001/77/EC and 2003/30/EC (OJEU no. L 140/16 of 5 June 2006).

transposing and often amplifying the provisions that figure in the international conventions. Although the directives from 1996 and 2003 were intended to be incentivizing, the texts adopted in 2009[25] reflect a change in political approach by combining climatic and economic issues. Also, with the aim to increase its energy security while reducing its dependence on oil and gas, the directive 2009/28/EC imposes on member states the adoption of measures for attaining, by 2020, three over-arching goals, labeled the "3 times 20" rule. First, it involves reducing by 20% greenhouse gas emissions compared to 1990 levels. Second, the share of renewable energy in the EU's primary energy consumption will be raised by 20%. Finally, energy consumption must be reduced by 20% by increasing energy efficiency. More recently, at the European Council of 24 October 2014, the European Union's heads of states, while defining the EU's action framework on the climate and energy objectives for 2030 in view of the Paris Conference of 2015, set the target of reducing their greenhouse gas emissions by 40%[26].

The first texts adopted in 1996, while affirming the objective of environmental protection, were mainly aimed at the creation of an domestic market for electricity. However, the texts adopted in 2003 connected renewable energies much more proactively to the objectives of competitiveness, security, supply and sustainability. Thus, it was affirmed that "a well-functioning domestic market in electricity should provide producers with the appropriate incentives for investing in new power generation, including electricity from renewable energy sources…". This new orientation is at the core of article 194 of the Treaty on the Functioning of the European Union, adopted in Lisbon, which relates to energy policy. It aims, among other things, to ensure the functioning of the energy market as well as the security of energy supply and to promote the effective development of renewable energies. It has since then been incorporated by the Commission, which notably included, in the "Energy Road Map 2050[27]", that pursuing the goal of "decarbonizing" the European Union must be carried out alongside the goals of competitiveness and security of energy supply. More directly relating to the marine energy sector, the Commission,

25 All of these texts, labeled a "Climate and Energy" package, are composed of four legislative tests (two directives, a regulation and a decision) each published in OJEU L 140 of 5 June 2009. See [THI 11].
26 Conclusions of the European Council of 24 October 2014, EUCO 169/14.
27 COM(2011) 885 of 15 December 2011.

while outlining its strategy for blue growth[28], presented this as one of the five growing fields of the blue economy that could help to boost employment in coastal zones. Since then, other initiatives from the Commission, such as the Communication on energy technologies and innovation[29] and the Action plan for the area[30], have recognized the importance of marine energy and aim to encourage collaborative research and development projects and cross-border cooperation to stimulate its development.

Faced with the requirement to put into practice its European commitments, France has gradually shaped an energy policy. The promotion of marine renewable energies relies on the use of programming and spatial planning instruments which are supported by the European legislation.

5.2.2. The planned and scheduled development of MRE

At sea, a space traditionally consecrated to freedom of movement, marine renewable energies represent a new practice which must be reconciled with other better known uses, in particular navigation, fishing, extraction of materials or underwater cable laying. The need to coordinate and reconcile human activities in the maritime and coastal space is primarily ensured by techniques of planning [QUI 13]. Sometimes described as a strategy, programme, scheme or pattern, the planning process consists of foreseeing individuals' potential actions in time and space. It is endowed with a legal framework which can be more or less binding, depending on its objectives and given principles.

In the maritime sphere, these planning techniques are part of the implementation of different public policy goals [BOI 13]: policies that involve energy, the environment, land-use, fishing, etc. Planning for marine and coastal areas uses a range of planning documents, some relating specifically to the sea. These should be used as a framework for the development of MRE (section 5.2.2.1). For now, however, it is the planning documents taken from energy law that set out the concrete conditions of this development (section 5.2.2.2).

---

28 COM(2012) 494 of 13 September 2012.
29 COM(2013) 253 of 2 May 2013.
30 COM(2013) 279 of 13 May 2013.

5.2.2.1. *Maritime and coastal spatial planning, a potential framework for the development of marine renewable energies*

Until the adoption of the Grenelle II law in 2012, the planning and management of marine activities was poorly covered by spatial and urban planning instruments (section 5.2.2.1.1). Since 2012, legislation has provided new instruments able to account for the MRE development, consistent with the principles of integrated management of coastal zones, and of maritime spatial planning (section 5.2.2.1.2).

5.2.2.1.1. The framework for planning of spatial and urban management

There are nowadays many texts for the planning of activities at sea, and many more to be developed. In addition to the instruments specific to spatial management policy, spatial planning documents belong to urban planning and environmental law. These are two areas of law that aim to control spatial planning and land use according to their respective objectives.

At first, marine zones were not really part of management issues. However, in 1983, French legislation provided for the creation of "*schémas de mise en valeur de la mer*[31]" (SMVM) [MES08][32]. The "*loi Littoral*[33]" of 1986 represented real progress in the consideration of coastal areas – mainly onshore areas – but did not push things forward for maritime zones. However, the provisions of this law, which amended the SMVM, reinforced the marine aspect of this instrument. After 30 years or so of experience, in mainland France only four SMVMs have been adopted and these represent quite small areas[34]. Despite several forecasting documents, in 2005, legislators decided to create a new category of SMVM within the territorial

---

31 lit: "Schemes for the enhancement of the sea".

32 SMVMs came after a prior experiment of "schémad' aptitudeetd' utilisation de la mer" (SAUM); lit: "schemes for aptitude and use of the sea".

33 lit: "Coastline Act".

34 These include the SMVM of the Thau Basin (1995), the Arcachon Basin (2004), the Gulf of Morbihan (2005) and Trégor-Goëlo (2007). It should be noted that Corsica and overseas departments and regions have their own management documents: the plan for sustainable management and development of Corsica and the schemes for regional management of overseas regions. These documents contain a part equivalent to the Coastal Planning Schemes and have precedence over other urban development planning documents in a compatibility report.

cohesion schemes. This, however, did not prompt any growth in the use of these marine planning instruments. The SMVM adopted did not concern, in practice, marine renewable energies; nor do they today represent documents which could be used on a relevant scale: the maritime coastline for the issues under consideration.

Other categories of spatial planning documents have also addressed marine areas[35]. The "directives territoriales d'aménagement" (DTA)[36], nowadays replaced by the "directives territoriales d'aménagementet de développement durable[37]" (DTADD)[38], aimed to transcribe France's vision for the management of the zones at stake. Several DTAs have involved estuaries where different marine activities are concentrated, such as the DTA of the Loire estuary[39]. The current DTADDs, although non-prescriptive, should nevertheless not be ignored. The French state can impose, using the "public interest project[40]" procedure, a development envisaged in an approved directive. In the future, the development of MRE could be facilitated if this procedure was made use of, especially where the electricity transport network's connection to the land causes problems.

With regard to planning legislation, administrative case law has recognized the principle that the territory of municipalities extends into the sea. Thus, planning documents apply to marine zones bordering the shores of the municipalities and therefore apply to the public maritime domain, natural as well as artificial. The "*plans locaux d'urbanisme*" (PLU)[41] should provide a zoning of these areas with appropriate rules [PRI 12]. For marine energy, the scope of this principle is quite limited. It will be shown below that MRE is not subject to the planning legislation procedures. It is, therefore, not

---

35 Spatial management policy also concerns large installations and networks with the adoption of the collective services schemes provided for by the land use planning and development act of 1999. Among these, the collective energy services scheme covers the development of renewable energy, but only briefly touches on installations at sea.

36 lit: "territorial development directives".

37 lit: "territorial directives for management and sustainable development".

38 Article L113-1 of the town planning code.

39 Examples include the TDD of the Alpes-Maritimes in 2003, the TDD of the Seine estuary in 2003, the TDD of the Loire estuary in 2006 and the TDD of Bouches-du-Rhônes in 2007.

40 lit: "public interest project".

41 lit: "local development plans".

required to conform to planning documents, particularly the PLU. However, the decisions on the approval of public maritime domain occupation must conform to the content of planning documents[42]. Municipalities can take into consideration activities related to MRE in their planning documents, either to assist their development or to limit it. Local planning rules unfavorable to certain MRE installations would nevertheless have little effect, in so far as when the French state backs a marine energy project, it has the ability to impose the necessary arrangements on the municipalities, using the "public interest project" procedure [BIL 13].

5.2.2.1.2. New instruments for sea and coastal planning

Beyond the SMVM experiment and other land development documents, planning of marine areas has established itself as an essential procedure in the development of activities at sea. The European Union has encouraged this planning, while legislation has introduced new instruments for this. The appearance of marine renewable energies in the maritime landscape and, in the short term, the installation of wind-turbines are one of the main reasons for needing spatial planning at sea. MRE is a new use for the sea which may change the current balance. The European Commission rightly asserts that "maritime zone planning is an important instrument for balancing the interests of different industries and to reach a sustainable use of marine resources, founded on an ecosystemic approach[43]". The European Union has since encouraged maritime spatial planning in the context of integrated maritime policy. In 2012, the Commission announced blue growth as a priority, meaning the creation of economic activities and employment linked to the sea and shoreline. Maritime spatial planning is, therefore, delineated as a means to increase the safety of investments in maritime activities[44]. More specifically, the European

---

42 Article L 2124-1 CGPPP "The decisions on use of public maritime domain take into account the destined use of the relevant areas and that of the neighboring land area, as well as the necessities to preserve the sites and landscapes of the shoreline and the biological resources; the decisions are thus in compliance especially with those concerning the neighboring land of public use".

43 Report of the Commission to the Council, the European parliament, the European Economic and Social Committee and Committee of the Regions, on the state of progress of the EU's integrated maritime policy, COM(2009) 540 final.

44 "Blue Growth opportunities for marine and maritime sustainable growth", COM(2012) 494 final.

Commission advised, in 2014, the integration of ocean energy into national programs for the planning of marine areas[45].

Previously, in 2008, the Marine Strategy Framework Directive has been an environmental pillar for integrated maritime policy[46]. According to this text, "the marine environment is a precious heritage that must be protected, preserved and, where practicable, restored with the ultimate aim of maintaining biodiversity and to maintain the diversity and dynamics of oceans and seas and to guarantee their cleanliness, healthy state and biological productivity". To do this, the States must adopt strategies for the marine environment, and action plans in order to achieve its good ecological status. In France, legislation has made provisions for the adaptation of the directive by creating "*plans d'action pour le milieu marin*[47]" (PAMM). These will be a subset of the strategic documents for the coastal areas, a new planning instrument given by the Grenelle II law of 12 July 2010[48].

The European Union has endeavored to incentivize maritime spatial planning. The European Commission produced a road map on this subject in 2008[49]. In particular, it has instigated different European programmes that encourage projects that foster cooperation or planning instruments. For example, the cross-border cooperation programme of ERDF[50] or the 7th framework programme for research and development (or FP7)[51]. The

---

45 "Blue Energy, Action needed to deliver on the potential of ocean energy in European seas and oceans by 2020 and beyond", COM(2014) 8 final of 20 January 2014.

46 Directive 2008/56/EC of the European Parliament and the Council of 17 June 2008 establishing a framework for Community action in the field of marine environmental policy (Marine Strategy Framework Directive) OJEU, L 164, 25 June 2008, 19.

47 lit: "action plans for the marine environment".

48 See articles 219-1 et seq. of the Environmental code under the law of 12 July 2010.

49 Roadmap for maritime spatial planning: achieving common principles in the EU, 25 November 2008, COM(2008) 791 final.

50 In the context of the European regional development fund (ERDF), the cross-border, transnational and inter-regional territorial cooperation programs can give support to marine planning projects, especially the operational projects "Two seas", "France (Channel)-England", "Italy-France Maritime", "Atlantic area", "Intereg IVC", etc. The programs very often prioritize the environment, territorial organization or the enhancement of the regions' natural and cultural assets. For example, the C-SCOPE project, financed by the "Two seas" program, carried out by the Dorset Coastal Forum and the coordination center for integrated coastal zone management in Belgium. This has enabled the creation of maritime planning documents.

51 Example: TransMasp (*Transboundary Maritime Spatial Planning*), Ghent University, 2011.

willingness to create maritime planning instruments has resulted in the draft directive for the establishment of a framework for maritime spatial planning and integrated coastal zone management. Thus, the pursuit of a more favorable environment for energy sector projects has instigated the adoption of the directive which sets out a framework for maritime spatial planning. However, the concerns of integrated coastal zone management have been dropped from this text[52]. The European Commission deemed that significant progress in maritime spatial planning has been made by the Member States. It underlined the importance of this process for the development of MRE and points to the need for cross-border cooperation for this[53]. The regional sea conventions, such as the OSPAR convention[54], are considered to be essential partners for the EU and have integrated the matter of maritime spatial planning into their field of action.

Before the European Union adopted an integrated maritime policy, international law established the integrated coastal zone management (ICZM) project. This aims to overcome the sectorial approach to coastal areas, integrating land and sea by applying an ecosystemic approach [VER 09]. The proposals of the Sea Grenelle, included in the law of 12 July 2010, have led to the adoption of a new body of rules for environmental law with the aim of integrated management of the sea and coastal areas. French legislation has thus introduced a compilation of the progress on integrated management of coastal areas and maritime spatial planning, while taking into account the marine strategy framework directive. The French state must adopt a national strategy for the sea and coastal areas, adapted to mainland coastal zones, using *"documents stratégiques de façade"* (DSF)[55] and *"documents stratégiques de bassin maritime[56]"* for overseas territories. DSF contains action plans for the marine environment in order to comply with the "Marine strategy" framework directive of 2008. It should be noted that the *"documents stratégiques de façade"* take pre-eminence over other planning instruments, especially terrestrial ones. This has been established by the

---

52 Directive 2014/89/EU of the European Parliament and the Council, establishing a framework for maritime spatial planning, OJEU L 257/135 of 28 August 2014.

53 Communication from the Commission to the European Parliament, the Council the Economic and European Social Committee and the Committee of the Regions "Maritime spatial planning in the EU – achievements and future development", COM(2010) 771 final.

54 The Convention of Paris, 22 September 1992, for the protection of the marine environment of the North-East Atlantic, known as the OSPAR convention, EMuT 992: 71.

55 lit. "Strategic documents on the shoreline".

56 lit. "Strategic documents on the marine basin", Article L219-1 of the Environmental code.

existence of a compatibility report in favor of DSF[57]. In the future run, these instruments should form the basis of MRE planning as they do for all other maritime activities, using an integrated approach.

The specificity of maritime law tends to preclude the application of many terrestrial rules and instruments. Legislation has given France a new planning framework specific to the sea. However, the desire to develop MRE has preceded the administration's ability to implement the "documents stratégiques de façade". This has led to the establishment of certain non-legislative documents: Offshore wind-turbine planning documents.

5.2.2.2. *Planning in the energy sector and its enforcement at sea*

The overall international and European objectives form the basis for the creation of the various programming and planning documents prescribed by European and national legislation. At European level, the 2009 directive on the promotion of the use of energy from renewable sources sets out that each Member state should adopt a national action plan for renewable energy that describes the means provided for its implementation. France's plan sets out the strategy devised for renewable energy without giving too many precisions on marine energies. The paragraphs that describe spatial planning relate to terrestrial cases, but are vague regarding the sea. Nevertheless, the document covers offshore wind-turbines, indicating the possibilities for the tender process and the purchasing prices of electricity produced in this way.

In France, the goals of the energy policy were defined by a "*loi de programme*[58]" of 13 July 2005, then, more specifically, by the Grenelle I law of 3 August 2009[59]. They have been compiled in a document named "the energy/climate plan" which sets out all the measures implemented by France in order to honor its international obligations and to achieve the goals of the Grenelle environment forum[60]. However, the development of renewable energies is legally dependant on the "*programmation pluriannuelle des investissements*" (PPI)[61]. This category of document was first prescribed by the law on the modernization and development of electric power public

---

57 Article L219-4 of the Environmental code.
58 lit. "Program law" – a law that describes the objectives fixed by the state in a given field.
59 Energy code, article L 101-1 et seq.
60 www.developpement-durable.gouv.fr/IMG/pdf/09003_PLAN_CLIMAT.pdf.
61 Articles L 121-3, L 141-1 and L 311-1 of the Energy code. lit: "multi-annual investment plan".

utilities (2000). The PPI sets out a framework for granting operation authorization or the tendering process[62]. The electricity production PPI accounts for the development of offshore wind-turbines. On the one hand, the document states the idea, also expressed in the national plan for renewable energy, of simplifying the procedures for the creation of installations for the production of electricity from renewable sources. On the other hand, it prescribes the creation of a forum for dialogue and seaboard planning.

Another planning aspect surrounding the development of MRE is the connection to the electricity transmission and distribution network. The transfer of produced energy often requires the creation of new network infrastructure. The waiting time for this can inhibit the development of renewable energies. Regional schemes for the connection of renewable energy to the network must help to anticipate the establishment of new transmission infrastructure for renewable energies[63]. More specifically, the national plan for renewable energy dictates that the tendering process "will help to share connections in the same area, in order to reduce the cost and the environmental impact[64]".

In reality, the planning documents for offshore wind-turbines are the result of the objectives stated in the energy programming documents and of a simple decision by the French government[65]. These documents are a pragmatic tool, established outside of any legal or regulatory framework, and can be classed in the category of sectorial planning documents. In 2009, the Ministry for Ecology decided to start collaborative planning, under the authority of the regions' prefects and in association with the competent maritime prefects, in order to define suitable areas for the development of offshore wind-turbines in the various regions with marine resources[66]. The prefects responsible for this co-ordination process carried out consultations

62 Law no. 2000-108 of 10 February 2000, *relative à la modernization et au développement du service public de l'électricité*; lit: "regarding the modernization and the development of electric power public utilities", partly repealed.
63 Article L 321-7 of the Energy code.
64   www.developpement-durable.gouv.fr/IMG/pdf/0825_plan_d_action_national_ENRversion_finale.pdf.
65 Letter of 5 March 2009 from the Ministry of Ecology, not published.
66 In the Mediterranean, the document was created for all the French Mediterranean territory, under the authority of the coordinating prefect, the prefect of the Provence-Alpes-Côte d'Azur region.

with the different stakeholders, in particular the developers of MRE projects and the representants of the fishers. The CETMEF[67] supplied the technical expertise necessary to establish a cartography of the different zones suitable for construction and the various constraints related to environmental protection: aviation and maritime safety as well as military, fishing and raw materials extraction activities, etc.

The adopted *"planification éolien en mer*[68]*"* documents have helped to identify some suitable areas. However, they have only managed to identify a limited number (as in Brittany), or have not helped to delineate these zones (as in the Mediterranean). The recipient government of the offshore wind-turbine planning documents has adopted the decisions taken on a decentralized level by proposing five different areas in Haute-Normandie, Basse-Normandie, Bretagne and Pays de la Loire in the tender of 5 July 2011. In a second tender, published 16 March 2013, the French government decided to again present the zone off the coast of Tréport, for which the previous tender did not bear fruit. They also presented a zone off the coast of Noirmoutier, identified by the planning document of the Pays de la Loire but not submitted for tender in 2011.

Planning of offshore wind-turbines is, in the end, an *ad hoc* process which is only just starting to come into energy law, but whose compatibility with environmental law has not been thoroughly deliberated. Indeed, it is straight away clear that certain planning documents effectively do not concern offshore energy, such as the *"schémas régionaux du climat, de l'air et de l'énergie*[69]*"* (SRCAE) and the *"schemas region aux éoliens*[70]*"* (SRE)[71]. Planning specifically for offshore wind-turbines was not provided for in the

---

67 *"Centre d'études techniques maritimes et fluviales"*; lit. "Centre for maritime and fluvial studies", the technical service at the center of the Ministry of Ecology and Sustainable Development, with national jurisdiction.
68 lit. "offshore wind turbine planning".
69 lit. "regional climate, air and energy schemes".
70 lit. "regional wind-turbine schemes".
71 It should nevertheless be noted when reading SRCAE adopted in seaboard regions that these documents sometimes touch on the potential of projects for developing ORE. The *"schemas région aux éoliens"* have been appended to the SRCAE by the Grenelle II law. They constitute the main planning documents for land-based wind-turbines after the loss of "wind-turbine development zones" as decided by the law no. 2013-312 of 15 April 2013, aiming to prepare the transition toward an efficient energy system and introducing various provisions for the pricing of water and wind-turbines, JO of 16 April 2013.

Environmental code. In addition, these regional documents of wind-turbine planning are not acknowledged by the regulatory texts which determine which documents should be submitted for impact evaluation, pursuant to the directive 2001/42/EC[72].

Drawing on international and European commitments, the creation of installations for the production of marine renewable energies is thus based on both programming instruments and planning tools. These define quantitative objectives and the areas for construction, while respecting the other uses of the sea and the protection of the marine environment. Access to this new market is governed by a specific legal framework which is gradually being developed.

5.3. The gradual development of a legal framework for ocean renewable energy

Access to the marine renewable energies market is subject to its own regulation and requires the fulfillment of complex administrative procedures which, in the words of professor Lavialle [LAV 13] "resembles an obstacle course" (section 5.3.1). In addition, the lack of an adapted legal framework leads to several legal and financial uncertainties (section 5.3.2).

5.3.1. *Access to the marine renewable energies market*

The economic activities in the field of marine renewable energies make up a market to which access is regulated (section 5.3.1.1). This requires administrative procedures which are as varied as they are complex (section 5.3.1.2).

5.3.1.1. *A regulated access*

Like the energy market which it is a part of, the MRE market is subject to regulation. The government must ensure competition in a market formerly held by one (or several) incumbent operator(s) in a monopoly (or

---

72 Directive 2001/42/EC of the European Parliament and the Council of 27 June 2001 on the assessment of the effects of certain plans and programs on the environment, OJEU L 197/30 of 21 July 2001.

oligopoly[73]). It must ensure that society's ongoing energy requirements are met while, more broadly, ensuring the security of the country's energy supply. This regulation is the result of the creation of a European energy market which is both open and competitive. The different adopted directives[74] impose, on the one hand, a complete separation of the activities of energy production, transmission, distribution and supply. On the other hand, they stipulate that energy production and supply be completely open to competition, while transmission and distribution remain under the control of the network operators carrying out public sector tasks. This opening to competition is now tightly governed by French public authorities and monitored especially by the regulatory authorities who are responsible for enforcing the rules and ensuring the proper functioning of the market.

In this context, France has organized a regulated access to the marine renewable energies market based on two preferred tools: the tender process [GUE 12] and the guaranteed feed-in tariff.

Article 8 of the law of 10 February[75], which transposes the European law[76], provides for two procedures: an authorization system and a tender procedure in cases where the initiatives of the market players would lead to too many differences compared to the programmed actions.

The authorization procedure applies when a project developer wishes to make its own application. It is described in articles L 311-1, L311-5 and L 311-6 of the Energy code. Authorization is supposed to benefit the construction of wind farms whose installed capacity is less than 30 MW. It is the Minister of Energy who gives this authorization and, where there are several competing applications, it is at his discretion to whom he will award the occupancy title. This procedure has not yet been implemented for two reasons. It is unreliable, since the government official retains the right to refuse, for whatever reason, to consider the concession project for the use of public maritime areas. Furthermore, the purchase price of electricity is not

---

73 A market in which there are just a few sellers in the face of a multitude of buyers.
74 See the European texts mentioned above and, in particular, the directive no. 2009/72/EC, the regulation (EC) no. 714/2009 and the regulation (EC) no. 713/2009.
75 Codified in article L 311-10 of the Energy code.
76 Directive no. 2009/72, replacing the identical provisions from the directive 96/92 and the directive 2003/54/EC that it repeals.

incentivizing enough to motivate developers to bid spontaneously for offshore wind-turbine projects[77].

In light of the lack of authorization requests, the tender process has been favored by the French government. The Minister of Energy, in line with the provisions of the decree of 4 December 2002 on this procedure for the electricity producing installations[78], chose to use this process, from 2004, for the construction of 21 wind-turbines, with a capacity of 105 MW, 6.5 km off the coast of Albâtre. Although the Energy Regulatory Commission expressed the view that the tender was inconclusive, the Minister accepted the application of the company CECA SAS Centrale ENERTRAG for the site of Veulettes-sur-Mer. It was authorized, by a decree of 13 October 2005, to operate an electricity production installation. Several appeals were thus lodged and finally the project was not implemented. Following the Grenelle II law, a new tender process was launched, this time for the construction and operation of five offshore wind-turbine installations[79]. Alongside this, a more recent tender process was published, 16 March 2013, for the construction of two new wind-farms with a combined capacity of 1,000 MW. These would be spread equally between two sites: one off the coast of Tréport, the other off the coast of Iles d'Yeu and Noirmoutier[80]. Furthermore, the government announced its plan to launch a tender for floating wind-farms in 2015, for the construction and operation of pilot tidal

77 In 2011, the price was 130 euros per megawatt. See: Minister of Ecology, Sustainable development and Transport/Minister of Economy, Finance and Industry, *"dossier de presse relatif au lancement du premier appel d'offres pour l'installation d'éoliennes en mer"*; lit: "press dossier on the launching of a first tender for the construction of off-shore wind-turbines" 2001, p. 9.
78 Decree no. 2002-1434 of 4 December 2002 on the tender procedure for electricity producing installations (JORF no. 288 of 11 December 2002, p. 20413), modified by the decree no. 2011-757 of 28 June 2011.
79 Published on the site of the CRE, 11 July 2011, it described throughout its 79 pages the technical, financial and administrative conditions that potential candidates must possess.
80 According to the specifications published 18 March 2013 on the site of the CRE, these two wind-farms complement the previous tender. The two projects, of Tréport and Noirmoutier, will each contain 80–100 turbines (giving a capacity of 480–500 MW) over areas of more than 100 km^2 in the Channel and the Atlantic ocean. The construction and gradual starting up of these two farms are expected for 2021–2023. As for the first tender, the CRE will give a score to each dossier based on the same criteria as the last time. However, a maximum qualifying price of 220€/MWh is fixed. The results have been published on 7 March 2014. The consortium made up of GDF SUEZ, EDP Renewables, Neoen Marine and AREVA won the two sites.

turbine farms in suitable areas in 2016 (in particular, the Raz Blanchard and the Raz Barfleur off the coast of Cotentin), and for wave energy installations in 2015/2016 (transforming wave and swell energy). The launch of the tenders is usually preceded by a call for expression of interest. This has the purpose of both consolidating the industrial sector and to encourage it to begin structuring itself [81]. Indeed, the tender procedure is used to achieve various public policy objectives, in particular for industrial development, environmental protection, sustainable development or spatial planning.

Nevertheless, its primary purpose is to organize a transparent and non-discriminatory competitive access to the marine renewable energies market and to enable the government to choose the most economically viable bid. As the Minister of Economy and Finance underlined in July 2011, the tender process is "a procedure of transparent and non-discriminatory specific allocation[82]". The government, therefore, undertakes the responsibility to guarantee the free access to this new market whose organization and monitoring it has been entrusted to the Energy Regulatory Commission.

The notice in the Official Journal of the European Union, the detailed content of the specifications document and the publication of the questions and answers of the prospective candidates on the site of the Energy Regulatory Commission are certainly for the sake of transparency. The requirement to introduce competition is reflected by the open opportunity for any candidate to make a bid and to receive equal treatment based on the specific criteria in the specifications document. According to the tender notice, published in the Official Journal of the European Union, the award criterion is for the most economically advantageous bid, judged according to

---

81 In 2009, ADEME launched a call for expression of interest (AMI) for marine energies in order to remove technological and non-technological barriers (environmental, economic impacts, etc.) in different marine energy sectors. Five projects were selected from this AMI (two tidal-turbine projects: SABELLA D10 and ORCA, two floating wind-turbine projects: VERTIWIND and WINFLO and a wave energy project: S3). A new AMI was launched on 13 March 2013 (closing date 31 October 2013). With 1275M€ of funds, its aim was to consolidate the ocean renewable energy sector and to cover the technological building blocks dedicated to the four sectors of marine tidal, wave, floating wind-turbine and thermal energy and the research demonstrators of wave energy plants. An AMI for the placement of tidal energy convertor demonstrators on the Raz Blanchard and Passage du Fromveur was launched on 1 October 2013 and the results were published on December 2014. See www.ademe.fr/.
82 JOAN 5 July 2011 – QE no. 101275 heading: "Marchés publics, passation"; lit: "Procurement contracts: awarding".

the subcriteria stated in the specifications document[83]. The criterion of price is given a weighting of 40%, the industrial aspect is also 40% and the socio-environmental criterion is 20%[84].

The right to bid is open to any person established on the territory of a Member state of the European Union, operating or wishing to construct and operate a production unit. However, the technical requirements, the industrial program and the required documentation imply that the bidder would have carried out long-term studies. Adding in the deadline of 6 months for candidates to make a bid, few operators were realistically able to make one. In the effort to structure, by the tenders, a promising industrial sector for marine renewable energies, the government, from the first tender, encouraged the operators to organize themselves to act jointly. It is thus possible to read on their official site that the Minister's decision on the four other areas of the 2011 tender "is based on the belief that a lasting industrial sector must involve several structuring operators, and that the industrial effort, and therefore the associated risk, should be spread across different operators, in order to ensure that the objectives of the environmental Grenelle are fulfilled in the long-term". Considering the technical requirements and the compulsory financial guarantees of the specification documents, it is clear that any application must involve an industrial partnership agreement. It is expected that the candidate has the technology and shows experience in development, installation, operation and maintenance in the fields of offshore wind energy, electricity production, or oil or gas extraction[85]. In other words, the candidate must be equipment manufacturer, energy operator and developer all at once.

In the end, bids were made by three companies, each making up a consortium, and only two won the tenders. The first winner was the "*société à actions simplifiées*" (SAS)[86] Eolien Maritime France (EMF) whose shareholders are the French energy operator EDF Energies nouvelles, and

---

83 According to the terms of the tender, this document is presumably the tender overview document available on www.cre.fr/.
84 See the specification documents (CDC) of the last two tenders: AO published 5 July 2011, CDC, chapter 5, point 5.2, Weighting of criteria, p. 30 and AO published 13 March 2013, CDC, point 5.2, p. 33. See www.cre.fr/.
85 Point 3.11 of the specifications document of the AO, published 5 July 2011, and point 4-2 of the AO published 13 March 2013. See www.cre.fr/.
86 lit. "simplified limited company".

Dong Energy Power, a Danish operator. This group involved several industrial partners, including the equipment manufacturer Alstom and the developer Nasset Wind. The second winner was the company Ailes Marines SAS, whose shareholders are Iberdrola, a Spanish energy operator, and the French developer EOLE-RES SA. They are in partnership, by consortium agreement, with the developer Neoen Res (affiliate of the Caisse des Dépôts), the equipment manufacturer AREVA, and with Technip, world leader in project management, engineering and construction for the energy sector. The third consortium, made up of the operator GDF SUEZ and the equipment manufacturer SIEMENS, had no successful bids. Indeed, the tender on the Tréport area was declared without result. According to the CRE, there was not sufficient competition for it. However, this site was again offered in a tender published 16 March 2013, and awarded in May 2014 to the consortium made up of GDF SUEZ, EDP Renewables, Neoen Marine and AREVA.

The tender process thus aims to select the candidates who will be authorized to construct and operate electricity production infrastructure from renewable energy sources. In return, they will profit from reselling the electricity they produce at the price they proposed in their bid[87].

As for all emerging energy sectors (nuclear, thermal and hydroelectric), wind electricity benefits from an incentivizing tariff to aid its development. The aim is to both encourage investment in these new technologies and to account for the very high implementation and operation costs of these new energy sources.

The principles that govern the purchasing obligation of energy from renewable sources are laid out in articles L 314-1 *et seq.* of the Energy code. In particular, article L314-7 states that the purchase price is designed to ensure normal profitability of the production of electricity from renewable sources. To achieve this, the price at which the energy distributor must buy the electricity is set, by decree, at a higher than market price. For wind energy, this measure is regulated by the decree of 17 November 2008.

---

87 Article 8 of the directive no. 96/92/EC of 19 December 1996 concerning common rules for the internal market in electricity (JOCE no. L 27/2 of 30 January 1997) now replaced by the directive no. 2009/72/EC (OJEU no. L 211/55 of 14 August 2009, p. 55). See law no. 2000-108 (JORF 11 February 200, p. 2143) and decree no. 2002-1434 (JORF no. 288 of 11 December 2002, p. 20413).

After mandatory consultations, the Energy Regulatory Commission, from 2001, issued a negative opinion on the prices set[88], stating that they gave a markedly excessive level of profitability on investments. In order to curb the prices, it encouraged the government to implement greater competition for offshore installations[89]. The purchase price set out by the decree of 2008 no longer applies for tenders. It is determined by the bidders[90] according to calculations described in the specifications documents[91].

Even though the proposed purchase price is only one of the three assessment criteria for the bid, the Energy Regulatory Commission (CRE) pays special attention to it. It is because of a particularly high purchase price that it recommended that the project at the Tréport site be abandoned. In particular, the purchase price proposed by EDF which it offered for the four other farms. However, as it stands, a price decline is far from being achieved. The CRE eliminated, during the procedure, the eliminatory character of a maximum price[92]. In the end, the government's expenditure for the *"contribution au service public de l'électricité"* (CSPE)[93] amounts to

---

88 Opinion of 30 October 2008 on *"projet d'arrêté fixant les conditions d'achat de l'électricité produite par les installations utilisant l'énergie mécanique du vent"*; lit. "the draft decree setting out the buying conditions for electricity produced by installations that use the mechanical energy of the wind", JORF no. 290 of 13 December 2008, replaced following its annulment by the Council of State, by the decree of 17 June 2014, JORF, no. 150 of 1 July 2014, p. 10827.

89 See the above-mentioned opinion of the CRE and the information bulletin of the CRE no. 5, June 2011.

90 Tréport, Fécamp, Courseulles-sur-Mer: min price 115 > max price 175 €/MWh and Saint-Brieuc and Saint-Nazaire: min price 140 > max price 200 €/MWh.

91 According to the specification documents, the price is the sum of two components: one "wind-energy project" component (POE) and one "connection to the transmission" component. The first component accounts for all the costs related to the study, implementation, operation and dismantlement of the installation, including the production units and electrical structures, as well as substations to the public networks for electricity transmission. The second component takes into account all the costs related to the study and creation of structures connecting between the public network of electricity transmission and the delivery point to this network (in accordance with decree no. 2007-1280 of 28 August 2008 on the coherence of connection and extension structures for connection to public electricity networks).

92 However, it reintroduced it for the tender launched for the Tréport and Noirmoutier sites. The specifications documents set an eliminatory maximum price of 220€/MWh for the electricity. The maximum price is also set at the median value, marked-up by 20%, "of the bids proposed by all of the applicants for the same site". Each applicant proposing a price over the maximum price "will be eliminated". See www.cre.fr.

93 lit. "contribution to public service charges for electricity".

1.2 billion euros per year. Without going into detail, this is because of an average purchase price, resulting from the tenders, very much higher than that prescribed by the decree of 17 November 2008. Yet, the legality of this decree was disputed by the Council of State on the grounds that the electricity purchasing mechanism that it establishes would constitute a state aid.

This is not a new issue since the Council of State has already considered it. Based on case law of the European Court in Luxembourg, it judged in a decree, given 21 May 2003 [CE 03, CE 08], that the purchase obligation arrangement, created by legislation to encourage the development of renewable energy, could not be considered as a system of state aid. However, since then, the case law has somewhat progressed. The European Court judged, in a decision given 17 July 2008 [CJE 08], that "when the governments mandate a company to manage a state resource, the purchase price mechanism should be seen as a state intervention through state resources", thus as an aid.

Following legal action for the annulment of the decree that set the purchase conditions of produced electricity, the Council of State, in a decree of 15 May 2012, decided to stay the proceedings and brought an action before the Court of Justice of the European Union on the following issue: "the purchase obligation mechanism of wind-generated electricity, should it be classed as state aid as defined by European law?". It should be clarified that state aid is not necessarily illegal. However, it cannot be granted without first going through a procedure of prior notification to the European Commission. To be classed as such, state aid depends on the presence of several criteria. The Council of State deemed that the Court of Justice of the European Union should give a ruling on the criterion of "conferral of a benefit". The Court unsurprisingly confirmed this in its decision of 19 December 2013. "The French mechanism of financing the purchase obligation of wind-generated electricity constitutes an intervention of the state, or through state resources, under article 107, paragraph 1 of the TFEU" [CJE 13]. Following the European Court's interpretation, the Council of State, in a decision of 28 May 2014 [CE 14], deemed that the purchase mechanism of electricity from renewable sources, established by the 2008 decree, was a state aid.

It would be wrong to think that this interpretation only applies to the guaranteed purchase price [BOI 14]. By fully offsetting EDF's (the obligatory buyer of electricity produced off-shore) surcharges for electricity public service, the French government participates, in the same terms, to the financing of this sector [GUE 13]. Following the annulment of the Council of State's 2008 decree, a new decree, of 17 June 2014, setting out the purchase conditions of electricity produced by wind energy installations, was published in the Official Journal on 1 July 2014. Prior to this, France had rectified the situation by notifying the 2008 mechanism to the Commission which, by a decision of 27 March 2014, validated it.

We believe that it is essential that the support scheme for the production of offshore wind-generated electricity follows this prior notification procedure. It is not unlikely that the mechanism might nevertheless escape classification as aid if it is shown that it constitutes a compensation for the obligation to participate in the operation of public electricity service and in the diversification of its supply sources[94]. Furthermore, drawing on §3 of article 107 aids for the facilitation of the development of certain activities, or for the contribution to certain environmental policy objectives may be declared as compatible with the common market[95].

While selection after the tender process gives the right to operate an offshore energy-producing installation, it does not at all predetermine the phase of alleviating the risks, then fulfilling the various obligatory administrative processes for authorization. Indeed, since the government does not have all the technical and environmental studies to guarantee the project's feasibility, the tender's specifications documents set out that the chosen applicant must produce, within 18 months of selection, a group of studies for the alleviation of the risks that might block the implementation of the project[96]. If the project cannot be implemented at the projected price, the competent Ministers have the right to withdraw the operation authorization and can progressively replace the purchase prices by feed-in premiums. This is the path that the government seems to wish to follow for its bill, currently being adopted, on "energy transition for green growth". It proposes the

---

94 Decision of the Commission no. 2007/580/EC of 24 April 2007 on the state aid scheme implemented by Slovenia in the framework of its legislation on qualified energy producers, OJEU L 219 of 24 August 2007, p. 9–24.

95 Guidelines on state aids for the environment and energy for the period 2014–2020, concerning state aids for environmental protection, OJEU C-82/1 of 1 April 2008.

96 §6 of the specifications document of the 2011 tender for offshore wind-farms. See www.cre.fr.

creation of an alternative mechanism called "additional remuneration". This would be a financial aid given in addition to the sale price for the market of electricity from renewable sources.

5.3.1.2. *A procedural complexity*

The creation of marine renewable energy installations requires that various procedures be followed. Relaying the European goal of simplification, France has gradually simplified the laws that apply to the installation and operation of marine renewable energies (section 5.3.1.2.1). However, this is a complex task which involves various procedural layers related to occupation of space, environmental protection and to public information and participation. Furthermore, different rules apply to the production installations and to the infrastructure for connection to the electricity transmission network (ETN). The rules vary as well depending on the technologies used (section 5.3.1.2.2). Finally, the applicable legal framework varies depending on the project area defined by the law of the sea. A project in an exclusive economic zone (EEZ) is now governed by a decree adopted on 10 July 2013[97] (section 5.3.1.2.3).

5.3.1.2.1. A desire to simplify the administrative procedures

The regime relevant to MRE is set apart from that of other renewable energies. It has its own specificities which are the result of the maritime character of these installations. The Grenelle II law outlined a procedural framework for this [BET 13]. The result is that the implementation of offshore energy installation projects requires, on the one hand, an "*autorisation domaniale*[98]" (or an authorization for occupation of an EEZ) and, on the other hand, an authorization under the water legislation. The first is only required for projects in the public maritime domain, up to 12 nautical miles from the shore. Beyond this, the regime for economically exclusive zones applies.

---

97 Decree no. 2013-611 of 10 July 2013 on "la réglementation applicable aux îlesartificielles, aux installations, aux ouvrages et à leurs installations connexessur le plateau continental et dans la zone économique et la zone de protection écologique ainsi qu'au tracé des câbles et pipelines sous-marins"; lit: "the regulations applicable to artificial islands, installations, constructions and their associated facilities on the continental shelf and in the economic area and the area of ecological protection as well as the cable route for underwater pipelines", JORF no. 0160 of 12 July 2013, p. 11622.
98 lit: "state authorization".

In the interests of simplification, legislators decided that the law concerning *"installations classées pour l'environement"* (ICPE)[99] does not apply to MRE[100]. Since then, the hazard studies prescribed by the legislation did not apply [BET 13]. However, the maritime prefect will have to adopt regulatory measures, for example, on fishing and navigation within the wind-farms, due to the potential risks involved. Similarly, legislation has canceled out the formalities and authorizations prescribed by the town planning code [BOR 09]. Installations for the production of energy from renewable sources, as well as the structures for their connection, are thus explicitly excluded from the scope of the town planning authorizations [BET 13].

5.3.1.2.2. Administrative hurdles on the way

For projects in inland seas and territorial waters, the key element of all the studied procedures is without doubt the concession to use the public maritime domain, provided for in the general code on public property (CGPPP). This code contains the principles from the coastlines act, the principles from the French *"loi littoral"* (1986) andand allows for the possibilities of authorization of occupation of the public maritime domain[101]. The concession demand procedure is determined by the decrees of the CGPPP on the concessions for the use of public domains outside of ports[102]. Authorization can be given only to allocate public domain dependencies for public use, public service or for operations in the public's interest. MRE projects quite definitely fall under this last category. However, the label of public service of MRE projects is less certain, electricity production – as opposed to transmission – not being a public service in principle[103].

According to the procedure, which is devolved to a departmental level, the MRE installation project organizer must submit a request to the prefect.

---

99 lit: "installations classified for the environment".
100 Decree no. 2011-384 of 23 August 2011, modifying the nomenclature of installations. The Council of State recognized the legality of the decree by the decree of 16 April 2012, *Volkswind France et Innovent*, req no. 353577.
101 We must take into account the provisions of the articles 2124-1 and 21242 from the CGPPP which determine the conditions of natural public domain occupation.
102 Art. R 2124-1 et seq. CGPPP.
103 EC, 29 April 2010, *Mr and Mrs Béligaud*, ° 323 179, *RFDA* 2010, p. 551, conclusions Guyomar M., note Melleray F.

This is then processed by the public maritime domain managing service[104]. In the context of tenders, the concession request comes after the Minister of Energy awards the contract for each of the delimited areas following the planning process. Thus, the prefect cannot give an authorization that would go against the results of the tenders. The winning bidder, who submits an application that respects the requirements of the tender and the regulations on public domain occupation and environmental constraints, should logically be granted the occupation authorization. In principle, the agreement of occupation of the natural public maritime domain does constitute an actual law. This raises the delicate question of the ownership of the installed structures, be it the base or the turbine in the case of offshore wind-turbines[105]. This hazy situation does help project financing. Finally, the convention project requires public consultation, in the form provided for by the Environmental code[106].

The second necessary authorization relates to the water legislation [BIL 13]. Because of their construction at sea, MRE installations can be subject to authorization or declaration according to *"police de l'eau[107]"*, whose provisions are set out in the Environmental code[108]. Offshore wind-turbine projects, as planned in the tenders, must obtain authorization from the prefect, pursuant to article R 214-1 of the same code, which specifies the nomenclature of operations subject to authorization[109].

The Environmental code requires that "public and private work projects, constructions and developments which, by their nature, their dimensions or by the location, may have a significant impact on the environment or on human health, must first be subject to an impact study[110]". Article R 122-1 of the Environmental code[111] submits all offshore energy production installations to an impact study. Prevention of harm to the environment is

---

104 Art. R 2124-1 et seq. CGPPP.
105 Art. L 2122-5 and R. 2124-9 CGPPP.
106 Art. R 2124-7 CGPPP.
107 lit: "Water policy".
108 Art. L 214-1 et seq. of the Environmental code.
109 The category of operations requiring an authorization is numbered 4.1.2.0 under heading V of the nomenclature: "Port area development works and other works carried out within the marine environment with a direct impact on this environment: 1° of a sum equal to or greater than 1 900 000 euros (A)".
110 Article L 122-1 of the Environmental code.
111 Specifically, the table appended to this article.

thus systematically ensured by an environmental evaluation which must accompany the application for occupation of public maritime domains. This evaluation must also be available to the public as part of the public consultation. When a Natura 2000 site is concerned, the MRE project developer must also carry out an evaluation of the Natura 2000 impacts.

Regarding information and participation, MRE projects must undergo public consultation. This obligation is the result of both the regulations on public maritime domain occupation and the regulations on water policy. Legislation requires that the project be subject to one single investigation[112].

Information and participation can also be carried out by a public debate. MRE projects fall under one of the operation categories for which the "*Commission nationale du débat public*" (CNDP)[113] must be referred to[114]. Following the results of the offshore wind-turbine tenders of 2011, the CNDP was referred to by the various winners. The CNDP decided to organize public debates for each site. Several findings were made from the organized debates, the reviews of the special public debate committees and the debates' summaries written up by the president of the CNDP[115]. In particular, they criticized the fact that the public debate took place so late, after the government had already chosen the sites. In addition, the lack of an impact study at this stage limited the publics' capability to discern the environmental issues especially, but not only, concerning the landscape, a focusing point of opponents to the various projects. Finally, the President of the CNDP recommended that the public consultation on each wind farm be carried out at the same time as the consultation on their connections.

In addition to the procedures related to public maritime domain occupation and environmental protection, MRE construction is also governed by the energy code. An operator wishing to produce electric energy from a marine source must obtain an authorization to produce electricity. This application is made to the Energy Minister[116]. However,

---

112 Article L 123-6 of the Environmental code.
113 lit. "National Commission for Public Debate".
114 Article 121-8 of the Environmental code.
115 See the documents on the offshore wind farms on the web page of the CNDP.
116 Article 2 of the decree no. 2000-877 of 7 September 2000, modified by the decree no. 2011-1893 of 14 December 2011 and articles L 311-5 and following the energy code. Installations of a power lower than a certain threshold (30 MW for wind-turbines) benefit from tacit approval since the decree no. 2011-1893, 18 February 2011.

most projects will only reach the tender stage, the only economically attractive procedure for the industry players. Knowledge of this procedure is, therefore, crucial in order to understand how to coordinate the various procedures for the implementation of MRE.

In reality, the use of tenders does not change the demands of the various legislations mentioned because of the principle of independence of laws. However, the tenders allow for the coordination of the procedures and give the prefect of the region a central role in the implementation of the administrative process, as the representative of the state. Indeed, the 2011 offshore wind farm tenders provided for a single referent, so that the candidates and, subsequently, the winning bidders could have effective communication with the administration. The tender's specifications document sets out various commitments for the winner: that they carry out, in good time and in a set order, the various regulatory obligations. From this, a timetable can be drawn up for the coordination of the procedures to follow. Some of the winner's commitments have also been added to the regulatory procedures. Thus, the risk control study must be submitted to the government in order to ensure that the project is feasible for the winner[117]. The chosen candidate must also provide performance guarantees which change over time.

An important aspect of the coordination of procedures is found in the distinction between the production installations and the electricity transmission infrastructure. For offshore wind-turbines, the winning bidder must distinguish between the project for the construction of the turbines and the electricity transmission works. Thus, in practice, there are two distinct groups of applications made by two distinct parties: the winning bidder for the turbines (or for another category of MRE), and ETN for the underwater and terrestrial cables, and the transformer. It should also be noted that the actor responsible for electricity transmission should, in fact, submit an occupation of public maritime domain concession application for the underwater cables from sea delivery point, as well as a construction permit application for these land facilities, for which other procedures should also be followed (for example, Natura 2000). On land, the structures for the connection of the transmission or distribution networks to marine installations using renewable energy benefit from an exemption of the building ban, prescribed by the French "*loi littoral*" (1986), in the 100 m

---

117 2011 offshore wind farm tender specifications document.

strip in front of the shore. They are similarly exempt of the rule that limits developments in remarkable landscapes[118].

5.3.1.2.3. The regime that applies to exclusive economic zones

Beyond 12 nautical miles, MRE installations are situated in the EEZ which may extend up to 200 nautical miles from the shore. For this space, the government no longer owns the sea floor and substratum. Therefore, public ownership law does not apply here. The current regime for EEZ activities is defined by the law of 16 July 1976 on the economic zone and the ecological protection zone of the coasts of French territories. Pursuant to this text and conforming to international law, France exercises sovereign rights for exploration and extraction of natural resources. In order to supervise installations in these maritime spaces, the government decided to create an authorization regime, by the decree of 10 July 2013 on the regulation applicable to artificial islands, installations, structures and their connected installations on the continental shelf and the economic zone and the ecological protection zone, as well as the laying of underwater cables and pipelines[119]. The regulatory text sets out an authorization procedure to submit to the maritime prefect and processed by the *"Direction départementale des territoires et de la mer"* (DDTM)[120]. The decree also provides for the possibility of a temporary authorization, of a length less than 2 years, for trial projects. The MRE operation application must contain information on the various aspects of the project: technical, financial, environmental, economic and social. The application must follow a number of formalities, such as financial guarantees and the impact study. In order to observe the principle of competition, a publicity procedure is prescribed, so long as the application contains the necessary guarantees. The procedure calls for the compilation of several administrative notices, as well as a public consultation by the competent authority. After the public consultation, prescribed by article 8 of the decree, the competent authority can make a definitive ruling, accounting for the concerns which it represents: in particular, navigation safety, the reversibility of the changes made to the natural environment and the sites, and the coexistence with the normal activities in the project area. Authorization entails aspects on the tracking of the project's impacts, as well as the measures and requirements of the project operator. It also sets out the conditions for the removal of the installations

---

118 Article L 146-4 III and article L 146-6 of the town planning code.
119 See decree no. 2013-611 of 10 July 2013, JORF no. 0160 of 12 July 2013, p. 11622.
120 lit. "Departmental board of territories and of the sea".

after operation and the necessary financial guarantees. Authorization cannot be given for a period greater than 30 years.

The regime that applies to EEZ does not exclude the implementation of rules regarding property rights. An MRE installation in an EEZ must be connected to the land. The electricity transmission infrastructure that crosses the public maritime domain is, therefore, subject to the rules of the CGPPP.

The other procedures under consideration do not change significantly for projects situated in the EEZ or on the continental plate, compared to ORE installations in territorial waters. The project coordinator must obtain authorization, under the energy code, and thus, in practice, be the winning bidder of a tender for an EEZ or continental shelf site. Similarly, environmental legislation must be respected. The authorization procedure, under the water legislation, and the procedures of the impact study, and the public consultation and debate also apply outside of territorial waters.

5.3.2. *A legal framework that leads to many uncertainties*

The legal framework of ocean renewable energy creates many legal and financial uncertainties. These are barriers to the development of these new energies and increase the risk of litigation.

5.3.2.1. *The legal uncertainties*

These legal uncertainties result from the candidate selection procedure, the reversibility and precarity of the granted authorizations and concessions, and the risk of legal action against the various decisions throughout the procedure.

5.3.2.1.1. The tender process

While the first article of the specifications document sets out that the winning bidder of a tender will be given operation authorization, the fulfillment of the project remains uncertain. The winners do have a right to operate an electricity plant, according to the Energy code. However, the absence in the texts of a time frame for obtaining this right[121] has raised

---

121 Neither article 311-1 of the energy code, nor decree no. 2002-1434 set out a maximum time for issuance of this authorization.

concerns with operators[122]. The decrees giving authorization to the winners of the four lots were quickly published in the end[123]. The winners can, therefore, expect to benefit from right to a feed-in tariff and, consequently, from the right to sign a contract for the purchase of the produced electricity with an "obligated" buyer, under the conditions of their bid and those set out by the specifications document.

The tender winners, who hold the right to operate an electricity production installation, cannot yet be sure of the project's implementation. Indeed, as the specifications document states, the awarding of the tender does not at all guarantee the successful outcome of the administrative authorizations procedures, which the winner must conduct under environmental law and for the occupation of public maritime domain.

Furthermore, according to the combined provisions of article 15 of decree no. 2002-1434 of 4 December 2002 on the tender procedure, the Minister has the right to not go through with it. This decision is based *a priori* on his discretionary power and therefore does not have to be justified. This is the solution that was taken for lot 1 (Tréport). The Minister declared the tender unsuccessful, in accordance with the CRE's recommendations[124].

A selected candidate unable to carry out the project must notify, stating their reasons, the Minister. They will then be subject to the sanctions set out in the specifications document. In this way, even if operation authorization has been given, the Minister, pursuant to article L 142-31 of the energy code, can impose a financial penalty, and withdraw or suspend the operation authorization. These sanctions do not prejudice the potential for redress of damages, of any nature, related to implementation of a new tender procedure.

---

122 Several questions have been made to the Energy Regulatory Commission (CRE) on the topic. See www.cre.fr/.

123 Three decrees of 18 April 2012 authorizing the company Eolien Maritime France to operate an electricity production installation off the shore of Fécamp, Saint-Nazaire and Courseulles-sur-Mer; and a decree of 18 April 2012 authorizing the company Ailes Marines SAS to operate an electricity production installation off the shore of the Saint-Brieuc commune, JORF of 28 April 2012, p. 7618.

124 It appears that the proposed purchase price was too high. The unfilled lot was however included in the second tender.

Furthermore, as stated by the CRE[125], the candidate is not granted exclusive rights over the allocated location. This means that other authorizations for the exploitation of natural resources, living or not, could be granted in the area of the lot.

The reversibility of authorizations and the precarity of the concessions

The authorizations requested as part of the implantation procedure can all be withdrawn [BET 13]. The authorization required for water, aquatic and marine environment protection can be "rescinded or modified without compensation from the government exercising its enforcement powers". This may be as a sanction to the operator or, in the case of a danger, for the safety of the public and the aquatic environment[126].

The concession to use the public maritime domain is also terminable, at any time, by the administrative judge on the request of the administration or, if the contract expressly provides for it, by the administration itself. This termination is delivered either for reasons of public interest, or as a sanction on the occupant for not complying with the requirements of the occupancy title, or for failing to respect the integrity, affectation or development of the area. Furthermore, various elements of the regime of concessions for the use of the public maritime domain are not adapted to the issues concerning the development of these new energies.

Therefore, the concessions granted to the operators do not constitute actual rights and do not give commercial ownership neither to the holders, nor to the contractors[127]. This raises the question of what rights do the concession holders have over the built structures. Without any specific information in the General code of public property, should it be considered that the base, fixed to the floor and substratum of the territorial sea, is subject to the rules that apply to the floor to which it is physically connected, and thus belongs to the government[128]? However, the rest of the wind turbine, constructed on land and then transported to the sea, is composed of a mast, blades and a nacelle which can be dismantled and are "simply" erected on the base. From a legal point of view, some deem that these may amount

---

125 Answer 71 to the questions asked and published on the CRE's site.
126 Article L 214-4-II and IIa of the Environmental code.
127 CGPPP: article R 2124-9.
128 Based on the doctrine of accession as defined by article 551, paragraph 1 of the civil code.

to personal property, and thus belong to the concession holder [BET 13]. Others deem that these elements, once erected on the base, are real property, due to their nature or intended use.

Another inadequacy of the regime of concessions results from measures which are essential for the conservation of the public maritime domain. The prefect can take these measures without any obligation to compensate the concession holder. "Therefore, a prefect's demand to dismantle a wind farm gives no right of compensation to the concession holder. Only a compensation clause of undepreciated investments can be added to the concession agreement, in case of termination for reasons of public interest" [LAB 10].

Appeals against the various decisions during the procedure

There is a proven risk of appeals [TER 13]. The CRE itself evokes this, by integrating delays for "litigation resolving", when outlining to the bidders the various elements of the procedure for access to market. We have already described the legal stalemate of the first construction project, off the coast of Albâtre, which forced the project leaders to abandon it. The second tender was also subject to appeals. From the 5 September 2011, the president of the "*Fédération environnement durable*[129]" lodged an informal appeal with the Ministers of Ecology and Industry for the withdrawal of the offshore wind turbine tender published 5 July 2011 in the Official Journal of the European Union. The mayor, the local residents' association and the fishers of Tréport, as well as the local residents' associations of Noirmoutier, La Baule, Saint-Nazaire, Saint-Brieuc, Arromanches and Veullette-Fécamp also affiliated themselves with this action. For the appeal petitioners, this tender "did not conform to the requirements of competition and certain provisions were illegal since they were based on a legally dubious decree". This appeal was rejected by the Nantes administrative court, 29 September 2011. Local associations for environmental or landscape protection, fishers, vacationers, and, of course, local residents can potentially contest the legality of administrative decisions for MRE implantation. The NIMBY[130] aspect is, therefore, not to be ignored, especially when it comes to owners of coastal property, concerned about the value of their assets which is partly linked to their views of the maritime landscape.

---

129 lit. "Sustainable environment Federation".
130 Acronym for not in my backyard.

5.3.2.2. *The financial uncertainties*

Financial uncertainties are also a reality. They involve the continuity of support systems, the applicable taxation, and the cost and relevance of dismantlement [MED 13].

5.3.2.2.1. A support system under suspension

Whether the constructions follow a tender or an authorization, all electricity consumers, since 2005, pay the price of support systems. A tax is added to their bill: *"contribution au service public de l'électricité"* (CSPE)[131]. Previously, these systems were financed by the *"Fonds du service public de la production d'électricité"* (FSPPE)[132], supplemented by contributions from the producers, suppliers and distributors.

However, as we have seen, following one association's appeal for the annulment of the decree of 17 November 2008, fixing the purchase price of wind-generated electricity, the Council of the State canceled this decree. They did so on the grounds that the mechanism of financing the obligation to purchase wind-generated electricity constituted a state aid, illegal without prior notice to the European Commission. In order to reassure operators, a new decree was made, 1 July 2014. This time, the government made sure to notify the Commission beforehand, who gave the authorization.

However, there seems little doubt that the support mechanism must evolve. Indeed, it is an unreliable economic system[133]. This is, in essence, the opinion of the Energy Regulatory Commission, which expressed an unfavorable opinion on the new decree. In its November 2012 newsletter, it stated that "the forecasted costs for 2013 increase by 43% (5.1 bn € vs. 3.6 bn €) compared to the costs for 2011". To this amount, "should be added...the adjustment for the year 2011, as well as the remaining costs from previous years. The costs for 2013 are therefore estimated at 7.2 bn €. Thus, in addition to the debt to EDF, the total contribution will only increase, given the development of alternative energies and increased aid to off-shore wind-turbines[134]". The question must inevitably be asked as to

---

131 lit. "contribution to public service charges for electricity".
132 lit. "public electricity service fund"
133 The sum of the CSPE for 2015 is 9.2 billion €. See www.cre.fr.
134 Report drawn up on behalf of the Commission for enquiry into the actual cost of electricity in order to determine the attribution to the different economic actors, report no. 667 of 11 July 2012.

what effect the cost of offshore renewable energy production will have on the customer. While the new guidelines, published by the European Commission, encourage the States to gradually move toward market mechanisms[135], the draft law on energy transition for green growth seems to account for it by proposing the creation of a new mechanism of "additional compensation", as mentioned above (section 5.3.1.1).

5.3.2.2.2. A tax system offering little incentives

Besides the usual contributions, a series of specific contributions, of ever-increasing value, is being added to all the costs borne by wind-farm operators. These include, for example, the annual tax on wind-turbines in interior waters or territorial seas, calculated on the installed power of each production unit. This can be revised each year[136]. Another example is the fee for occupation of the public maritime domain, calculated on the number of turbines and the length of the connection. Electricity production installations, situated in interior waters or territorial waters (article 1519D of the General tax code), using the mechanical hydraulic energy of currents, incur the "imposition forfaitaire sur les entreprises de réseaux"(IFER)[137], calculated on the number of MW.

However, a lightened and adapted tax system, which accounts for the development and operation costs caused by the peculiarities of ORE, would encourage investment in these sectors.

5.3.2.2.3. Dismantling

The specifications document of the tender obliges the winning bidders to dismantle the installations at the end of the concession. Such a process, planned for after 30 years of operation[138], appears somewhat questionable in light of the limited number of areas suitable for offshore development [SAN 09] and shows little willingness to develop offshore wind energy in the long term. Furthermore, there is a lack of information on the environmental impact of such dismantlement. The base of the turbine can,

---

135 Communication no. 2014/C 200/01 of 28 June 2014 on the guidelines on State aid for environmental protection and energy for the period 2014–2020, OJEU no. C 200 of 28 June 2014.
136 Article 1519B of the general tax code, modified in 2013 by decree no. 2013-463 of 3 June which raises the value of the tax to 14 480 euros per installed megawatt.
137 lit. "flat-rate tax on network businesses".
138 Article R 2124-1 CGPPP.

over time, become home to a true ecosystem, which would be destroyed or damaged by such a step [ROC 07].

To the concept of "dismantlement" provided for in the Environmental code[139], it would without doubt have been preferable to refer to, in the tender, "the reconditioning, restoration or rehabilitation of the sites at the end of the title or use", as is given in the General code on public property[140]. Indeed, the latter provisions appear not to oblige the operators to dismantle their installations. They open the possibility to maintain the site as "brown land", by keeping the structures' bases, while waiting for a change of use or an assignment of the base, so long as the interests referred to in articles L 211-1 and L 511-1 of the Environmental code are respected (respectively, the sustainable and balanced management of water resources, and public health, safety and hygiene with protection of the environment) [CAR 10]. While reconditioning "consists (simply) of carrying out works to remove traces of the operation and to aid the reintegration of the plots into their area and, more generally, into the environment", dismantling "requires the disassembly and removal of superstructures and machines, including the foundations and the delivery substation[141]". Decree no. 2013-611 on EEZ gives a different perspective on the matter. It describes the removal of the installations, based on the authorization procedure given in article 17 of the above decree. The result is that, at the end of the concession, the operator will prepare a report which gives a detailed programme for the removal operations. This will be submitted for approval by the competent authority, who will rule on the compatibility of this programme with the activities exercised in the area. It is interesting to note that the competent authority may decide to keep certain elements if they benefit the ecosystems and do not prejudice safe navigation.

5.4. Conclusion

The procedure for the installation of offshore wind-turbines is rather unique. Despite efforts to simplify the process, it remains complex and uncertain. We can only agree with the opinion of professor Lavialle that

---

139 Article L. 553-3 of the Environmental code.
140 Articles R 2124-2-8° and R 2124-8 CGPPP.
141 For these definitions, see the *Guide de l'étuded'impactsurl'environnement des parcséoliens,* distributed by the services of the Ministry of Ecology, Energy, Sustainable Development and the Sea.

"where one might have envisaged, given their particularity compared to onshore structures, that an original legal framework might have been introduced, public authorities have in fact been satisfied to retain, as is, all the pre-existing procedural elements that apply to the building of structures in the marine environment, hence the pile of constraining and redundant provisions" [LAV 13]. It is, therefore, desirable that the next tenders for the construction of offshore renewable energy installations are accompanied by efforts to adapt the French legal framework. The simplification and securing of procedures does not preclude a change in legislation.

5.5. Bibliography

Texts and case law

[CE 03] CE, Union nationale des industries utilisatrices d'énergie, UNIDEN, no. 237466, 21 May 2003.

[CE 08] CE, Association Vent de Colère, no. 297723, 6 August 2008.

[CE 14] CE, Association Vent de Colère, no. 324882, 28 May 2014.

[CJE 64] CJCE, Costa c/. ENEL, aff. 6/64, rec. 1141, 15 July 1964.

[CJE 94] CJCE, Commune d'Almelo et autres, aff.C 393/92, rec. I-1508, 23 April 1994.

[CJE 08] CJUE, Essent Netwerk Noord BV, affaire C-206/06, 17 July 2008.

[CJE 13] CJUE, Association Vent De Colère! Fédération nationale et autres contre ministre de l'Ecologie, du Développement durable, des Transports et du Logement et ministre de l'Economie, des Finances et de l'Industrie, aff.C-262/12, 19 December 2013.

[JOC 94] JOCE, Convention des Nations unies sur le changement climatique, JOCE L.33 du 7 February 1994.

[JOC 02] JOCE, Kyoto Protocol, adopted 11 December 1997, JOCE L.130 du 15 May 2002.

[JOR 10] JORF, Loi no. 2009-967 du 3 August 2009 de programmation relative à la mise en œuvre du Grenelle de l'environnement, JORF no.179 du 5 August 2009, p. 13031, et loi no. 2010-788 du 12 July 2010 portant engagement national pour l'environnement, JORF no.160 du 13 July 2010, p. 12905.

Report

[BOY 13] Boye H., Caquot E., Clément P. *et al.*, Rapport de la mission d'études sur les énergies marines renouvelables, ministère de l'Ecologie, du Développement durable et de l'Energie et ministère du Redressement productif, disponible sur www.developpement-durable.gouv.fr/IMG/, March 2013.

Doctrine

[BET 13] Bettio N., "La procédure d'implantation des éoliennes offshore en droit français", in Gueguen-Hallouet G., Levrel H. (eds), *Les énergies marines renouvelables – Enjeux juridiques et socio-économiques*, Actes du colloque de Brest des 10 et 11 October 2012, Pedone, Paris, pp. 73–92, October 2013.

[BIL 13] Billet P., "L'exploitation de l'énergie marine au risque du droit de l'environnement", *Les énergies marines renouvelables – Enjeux juridiques et socio-économiques*, Actes du colloque de Brest des 10 et 11 octobre 2012, Pedone, Paris, pp. 41–52, October 2013.

[BOI 13a] Boillet N., "La planification des énergies marines renouvelables en droit français", in Gueguen-Hallouet G., Levrel H. (eds), *Les énergies marines renouvelables – Enjeux juridiques et socio-économiques*, Actes du colloque de Brest des 10 et 11 octobre 2012, Pedone, Paris, pp. 53–70, October 2013.

[BOI 13b] Boiteau C., "Mise en perspective des fondements internationaux et européens du droit de l'énergie renouvelable", in Gueguen-Hallouet G., Levrel H. (eds), *Les énergies marines renouvelables – Enjeux juridiques et socio-économiques*, Actes du colloque de Brest des 10 et 11 octobre 2012, Pedone, Paris, pp. 7–24, October 2013.

[BOI 14] Boiteau C., "Mécanismes de soutien à la production des énergies renouvelables et le droit des aides d'Etat: le cas de l'éolien", in Boiteau C. (ed.), *Energies renouvelables et marché intérieur*, Chapter 1, Bruylant, Paris, pp. 1–20, 2014.

[BOR 09] Bordereaux L., Braud X., *Droit du littoral*, Gualino, Paris, 2009.

[BOR 12] Bordereaux L., Roche C., "Du droit du littoral au droit de la mer: Quelques questions autour des énergies renouvelables", *Droit maritime français*, p. 1038, December 2012.

[CAR 10] Carpentier A., "Eoliennes et installations classées, acte I", *Actualité Juridique Droit Administratif (AJDA)*, p. 2030, 2010.

[COU 11] COULOMBIE H., LE MARCHAND C., *Droit du littoral et de la montagne*, LITEC, Paris, 2011.

[GRA 11] GRARD L., "Les racines européennes de la nouvelle organisation française du marché de l'électricité", *Revue Europe*, pp. 5–10, March 2011.

[GUE 12] GUEGUEN-HALLOUET G., BOILLET N., "L'appel d'offres éolien en mer", *JCP. A*, no. 40, étude no. 2320, 8 October 2012.

[GUE 13] GUEGUEN-HALLOUET G., "L'appel d'offres éolien offshore à l'épreuve des règles européennes de concurrence", in GUEGUEN-HALLOUET G., LEVREL H. (eds), *Les énergies marines renouvelables – Enjeux juridiques et socio-économiques*, Actes du colloque de Brest des 10 et 11 octobre 2012, Pedone, Paris, pp. 93–106, October 2013.

[LAB 10] LA BOUILLERIE P., MARTOR B., "Projets éoliens offshore: un nouveau souffle électrique en haute-mer", *JCP E.*, no. 16, p. 1394, April 2010.

[LAV 13] LAVIALLE C., "Implantation et espaces marins-Rapport de synthèse", in GUEGUEN-HALLOUET G., LEVREL H. (eds), *Les énergies marines renouvelables – Enjeux juridiques et socio-économiques*, Actes du colloque de Brest des 10 et 11 octobre 2012, Pedone, Paris, pp. 143–154, October 2013.

[MED 13] MEDDEB M., "Approche comparative des dispositifs de soutien aux énergies marines renouvelables", in GUEGUEN-HALLOUET G., LEVREL H. (eds), *Les énergies marines renouvelables – Enjeux juridiques et socio-économiques*, Actes du colloque de Brest des 10 et 11 octobre 2012, Pedone, Paris, pp. 283–294, October 2013.

[MES 08] MESNARD A.H., "Schémas de mise en valeur de la mer", *Droits maritimes*, no. 521, Dalloz, Paris, 2008.

[MON 14] MONACO A., PROUZET P., *Value and Economy of Marine Resources*, ISTE, London, and John Wiley & Sons, New York, 2014.

[PAI 14] PAILLARD M., MULTON B., BŒUF M., "Marine renewable energies", in MONACO A., PROUZET P. (eds), *Development of Marine Resources*, ISTE, London, and John Wiley & Sons, New York, 2014.

[PRI 12] PRIEUR L., "Le PLU littoral", *Les Cahiers du GRIDAUH*, no. 23, 2012.

[QIU 13] QIU W., JONES P.J.S., "The emerging policy landscape for marine spatial planning in Europe", *Marine Policy*, no. 39, pp. 182–190, 2013.

[ROC 07] ROCHE C., "Et pourtant, elles tournent: la réglementation applicable aux éoliennes *offshore*", *AJDA*, p. 1785, 2007.

[ROC 12] ROCHARD A., "L'appel d'offre éolien offshore", *Contrats publics*, no. 121, p. 46, March 2012.

[SAN 09] SANDRIN-DEFORGE A., "Aperçu sur la réglementation applicable aux projets éoliens en mer", *BDEI*, no. 21, p. 37, June 2009.

[TER 13] TERNEYRE P., "Le droit applicable au marché des énergies marines renouvelables", in GUEGUEN-HALLOUET G., LEVREL H. (eds), *Les énergies marines renouvelables – Enjeux juridiques et socio-économiques*, Actes du colloque de Brest des 10 et 11 octobre 2012, Pedone, Paris, pp. 255–260, October 2013.

[THI 11] THIEFFRY P., "La politique de l'énergie", *Droit de l'environnement de l'Union européenne*, 2nd ed., Bruylant, Paris, pp. 1156–1206, 2011.

[VER 09] "Quelle stratégie européenne pour la gestion intégrée des zones côtières", *VertigO – la revue électronique en sciences de l'environnement* [en ligne], Hors-série, 5 May 2009.

Socio-economic Evaluation of Marine Protected Areas

6.1. Introduction

Marine protected areas (MPAs) are a more recent creation than nature reserves on land. They appeared toward the middle of the 20th Century[1] and have seen a rapid development in the last 30 years: from 430 in 1985 [DES 86], the total number of MPAs throughout the world increased to 1,300 10 years later, reaching 6,500 in 2014. This represents 2.1% of the total surface area of the world's seas (www.mpatlas.org).

In France, the national park of Port-Cros, created in 1965, constituted the first MPA. In 2013, according to data published by the Agency of marine protected areas, 392 MPAs were counted in waters under French jurisdiction, of which 290 are situated off the mainland, and 102 are overseas (www.aires-marines.fr). The fraction of waters under French jurisdiction with MPA status, less than 0.1% at the beginning of the 21st Century, has since increased rapidly to reach 3.8% in 2013. The *"Grenelle de l'environnement"* (2007) and the *"Grenelle de la mer"* (2009) outlined the

Chapter written by Frédérique ALBAN, Jean BONCOEUR and Jean-Baptiste MARRE.
This study was funded as a part of the BUFFER European research program dedicated to multipurpose MPAs ("Partially protected areas as buffers to increase the linked socio-ecological resilience", ERANET BIODIVERSA 2013-2015).
1 Fort Jefferson National Monument, created in Florida in 1935, is often considered as the oldest MPA. It became the Dry Tortugas National Park in 1992.

ambitious target to confer MPA status to 20% of the waters under French jurisdiction between now and 2020.

MPAs are very diverse, varying in terms of location, size, the characteristics of the natural, economic, social and cultural environment in which they exist, the objectives that they pursue, their legal status, governance, the protective measures employed and the effectiveness of these measures[2].

In France, the law of 14 April 2006 distinguishes six categories of MPA, defined by their legal status: national parks with a marine section, natural reserves with a marine section, prefectoral orders for the protection of biotopes, natural marine parks, Natura 2000[3] sites with a marine section and the section of maritime public domains entrusted to Coastline Conservation[4]. The Agency of protected marine areas distinguishes eight potential goals for the creation of these MPAs:

– F1. the healthy state of listed and heritage species and habitats or those that deserve to be (rare species, threatened species);

– F2. the healthy state of unlisted species and habitats (exploited species, very locally abundant species giving biogeographical responsibility to the host site);

– F3. the yield of key economic functions (spawning grounds, nurseries, productivity, resting, food supply and migration);

– F4. the healthy state of marine waters;

– F5. the sustainable use of resources;

– F6. the sustainable development of usages;

– F7. the maintenance of maritime cultural heritage;

– F8. added value (social, economic, scientific and educational).

---

2 A large proportion of MPAs found across the world are "paper parks", meaning structures that only exist on paper.
3 Natura 2000 is a network of EU natural or semi-natural sites with a high heritage value, due to their remarkable flora and fauna.
4 This list is detailed in the prefectural order of 3 June 2011, which introduced nine new categories of MPA, mainly linked to international conventions.

These objectives vary according to the category of MPA in question.

Even though objectives of a socioeconomic nature are present in the list created by the Agency (see in particular F5, F6 and F8), the primary objective of MPAs is not economic and social development, but the protection of the environment. This characteristic is clearly reflected by the definitions most frequently given for the notion of an MPA (see [GAR 13] Chapter 2 for a review).

Categories of marine protected area as per the law of 14 April 2006	Potential objectives							
	F1	F2	F3	F4	F5	F6	F7	F8
Natural reserve with a maritime section	×	×	×					×
Natura 2000 site at sea	×							
National park with a maritime section	×	×	×	×	×	×	×	×
Natural marine park	×	×	×	×	×	×	×	×
Marine sections of MPD being managed by Coastal conservation	×	×	×			×	×	×
Biotope protection order with a maritime section	×							
	Source: www.aires-marines.fr							

Table 6.1. *Objectives assigned to French MPA*

The French law of 14 April 2006, which introduced into national law the notion of MPAs, does not give a definition of this notion. However, the Agency of marine protected areas created by the same law MPAs as "a specific area at sea that meets the objectives of protecting nature over the long-term" (www.aires-marines.fr)[5]. The most widely accepted definition is given by the International Union for the Conservation of Nature (IUCN): "an area of intertidal or subtidal terrain, together with its overlying water and associated flora, fauna, historical and cultural features, which has been reserved by law or other effective means to protect part or all of the enclosed environment" (www.iucn.org)[6].

---

5 On the same website, the notion of an MPA is also defined as a "a defined space that meets the objective of protecting nature over the long term, not exclusive from controlled economic development, for which management measures are defined and implemented" (*Ibid.*).
6 This definition hides the fact that numerous areas across the world defined as MPAs include terrestrial areas.

Under these conditions, the socio-economic evaluation of MPAs may seem to be of minor concern, the subject matters of MPAs relating essentially to field of natural sciences. However, limiting ourselves to this vision may impede the implementation of public policy for the development of MPAs.

MPAs may indeed be characterized as "an investment by society in the conservation of its natural capital" (Alban *et al.* in [CLA 11] Chapter 9). This investment may have various incentives: the protection of the ecosystem may be an objective in itself, but it also constitutes an intermediate objective in order to safeguard the sustainability of certain usages of the ecosystem (fishing, tourism, etc.). Each of these incentives can be grouped together under the general term of "protection of ecosystem services", the services provided by the natural capital of the MPA's ecosystem[7].

Alongside the expected positive effects (benefits), the implementation of such an investment inevitably involves negative effects (costs). These do not only include the management and surveillance costs of the protected area, which may be considered as running costs. However, they also include the less explicit, but just as real, costs which economists call opportunity costs. These are the result of the fact that the investment in question ties up scarce resources which could have otherwise been profitably employed in some other way. For MPAs, the main source of opportunity costs arises from the restrictions, in the name of environmental protection, which are imposed on the users of the area and its resources (restrictions or a total ban on fishing, for example)[8]. These costs can be high, such that it is important for policy makers to determine how best to share these out among the stakeholders, and to what extent the benefits

---

7 The term "ecosystem services" was popularized at the start of the 2000s by the *Millennium Ecosystem Assessment*, a study commissioned by the Secretary General of the UN in 2001, which resulted in a series of reports published between 2003 and 2005 (www.millenniumassessment.org). It directly refers to the economic definition of the term "capital", seen as a durable object (material or immaterial, natural or artificial) that creates, over time, "services", meaning effects that are positively appreciated by humans. In principle, the value of a capital is equal to the sum of the present values of the services that it gives over its entire lifespan (which may be infinite). In this vein, the "total economic value" of an ecosystem (a concept popularized in particular by [COS 97]) is defined as the sum of the present values, in principal over an infinite amount of time, of all the ecosystem services that it creates (for more details on discounting see below, section 6.2.1).

8 The concept of an MPA is sometimes confused with that of a fishing reserve. In reality, the areas in which fishing is banned make up only a fraction of the surface area of the MPA: in 2014, these areas covered 0.89% of the total surface area of the world's seas, 43% of the total marine surface area of MPAs (www.mpatlas.org).

compensate them. Their underestimation, or omission, is often a cause of failure for MPA projects.

Socioeconomic evaluation of MPAs aims to identify and measure the social costs and benefits created by the MPA. Thus, an overall assessment and its breakdown within the society can be determined. This evaluation may occur *ex ante* (as a decision aid for the creation of the MPA) or *ex post* (to monitor the implementation and to define potential corrective action). Evaluation becomes all the more necessary as the number of MPA projects increases, while the means to implement them are often becoming scarce. Similarly, the search for sustainable sources of funding is often behind the demand for an evaluation: it entails characterizing the ability to contribute of those who are set to gain the most from the MPA.

In this chapter, we will first present the main tools for the socioeconomic evaluation of MPAs. We will then discuss some of the problems faced during their implementation and the adaptations often used in the attempt to resolve these problems. Finally, using recent studies we will try to identify the practical role of socioeconomic evaluation of MPAs.

6.2. Methods

The evaluation of the positive and negative effects of an MPA on society comes under the field called project analysis. After introducing the philosophies of the two major families of project analysis methods, we will present the two types of tools for their implementation in the case of MPAs: the techniques for the evaluation of non-market values and bioeconomic models.

6.2.1. *Project analysis methods*

The aim of project analysis methods is to enable policy makers to determine the benefit of a project (for example, the creation of an MPA) on two levels: its effectiveness, meaning the project's ability to create a surplus of social welfare; and also its equity, which relates to the way in which the positive and negative effects are distributed throughout society.

Project analysis methods may be divided into two broad categories: cost-benefit analyses (CBA) and multi-criteria analysis (MCA). These two categories are differentiated by the metrics used to quantify the effects of a project: while CBA use one single metric (money), MCA use metrics that vary depending on the effects being considered.

CBA started to develop in the United States in the first half of the 20th Century for the programming of public works. Nowadays, environmental management is an important application of the CBA (see, for example, [PEA 06]). This method is often advised [HOA 95] and sometimes implemented for the economic evaluation of MPAs (see, for example, [CLE 10, MAN 13] and [PAS 11]). In CBA, the evaluation of the effectiveness of a project is based on a single criterion, intended to summarize all of the benefits and costs, each expressed in monetary terms. This criterion is conventionally the net present value (NPV), defined as the algebraic sum of the present costs and benefits[9]:

(where I represents the cost of the initial investment, n represents the project's lifespan, i represents the discount rate, A_t and C_t represent the total monetary values of the benefits and costs throughout year t).

One variant is the internal rate of return (IRR), defined as the discount rate which brings the NPV of the project to zero[10]. Table 6.2 presents the estimated IRRs of a few MPAs which have recently undergone CBAs on the request of the donors who financed their creation (except for the first one, note the very high rates which suggest a remarkable effectiveness of the MPAs whose effects they are meant to synthesize)[11].

---

9 Discounting is a technique that allows economic flows staggered over time to be compared. If an operation Y_t (for example, a revenue) must occur in t years and if, from now until then, it is possible to loan or borrow with an annual rate of interest of i, the present value, or current equivalent of Y_t is $Y_0 = Y_t(1 +i)^{-t}$. Indeed, investing Y_0 over t years at a rate of i (with interest capitalization) would give, after t years, an acquired value of $Y_0(1 + i)^t = Y_t$.

10 In the simplest case, the NPV of a project is a monotonously decreasing function of the rate i used in the time discounting of the future effects of this project (this is a case when $A_t > C_t \ \forall t \geq 1$). It becomes negative when the rate exceeds a certain threshold i^*, which is by definition the IRR of the project. It is considered that the higher this threshold, the greater the return on the project. Compared to NPV, the IRR criterion has the advantage of not depending on a discount rate fixed *a priori*.

11 As a comparison, the average rate of return before tax of Europe's manufacturing industry was around 6–8% before the economic crisis that erupted in 2008 (BACH European database).

MPA assessed	Estimated IRR (central scenario)
MnaziBay (Tanzania)*	3%
Bamboung (Senegal)*	26%
Quirimbas (Mozambique)*	31%
Soufrière (Saint Lucia)*	57%
Emua, Laonamora, Piliura, Unakap, Worasifiu (Vanuatu)**	41%

Table 6.2. *Summarized results of benefit-cost analyses of MPAs (sources: * [CLE 10]; ** [PAS 11]. Reproduced from [GAR 13] Chapter 7)*

MCAs have more recent origins than CBAs. Developed in the 1960s, the ELECTRE method [ROY 68] is generally considered as the prototype for MCAs[12]. Since then, these methods have been used for a very wide range of applications, particularly in the field of environmental management (see, for example, [CGD 14]), and several of these involve MPAs (see, for example, [BRO 01] and [VIL 02]).

Unlike the CBA, MCA does not call for a single metric to assess the different effects of a project: each effect is measured using its most suitable unit (monetary, physical units and scores). Because of the heterogeneity of the evaluation criteria, it is necessary to define an algorithm in order to rank the projects when their performances in each of the different criteria are not consistent (for example, when project 1 has a better performance than project 2 in criterion A, but a worse performance in criterion B). This algorithm is typical of the method in question. It is usually based on the assignment of weighting coefficients to the different criteria which reflect their importance to the decision-maker[13]. Box 6.1 shows a simple example of a MCA applied to the socioeconomic evaluation of three MPAs [BON 10].

As a part of the AMPHORE project, the socioeconomic performances of three MPAs (the community MPA of Bamboung in Senegal, the national park of Banc d'Arguin in Mauritania and the national park of Port-Cros in France) were evaluated using an MCA that took into account five criteria determined by field surveys and expressed as scores on a five point scale (0 for the lowest and 4 for the highest). A panel of experts was consulted for the weighting of each of the criteria and the ranking given by the individual members was aggregated using the Borda count method. In this very simple method, each person participating in the consultation

---

12 For a presentation of MCAs with case studies, see, for example, [ROY 93] or [DOD 09].
13 This is, however, not the case for ELECTRE, which uses an outranking algorithm.

ranks the *n* criteria in order of importance (here *n* = 5). While reading the ballots, the criterion ranked in first position is given a number of points equal to *n*, *n*-1 if it is ranked in second position, etc. Afterward, the total number of points obtained by each criterion is counted (this total is called the "Borda score") and the indicators are classed according to the totals obtained by each one. After standardization, the Borda scores for each criterion can be used as weighting coefficients.

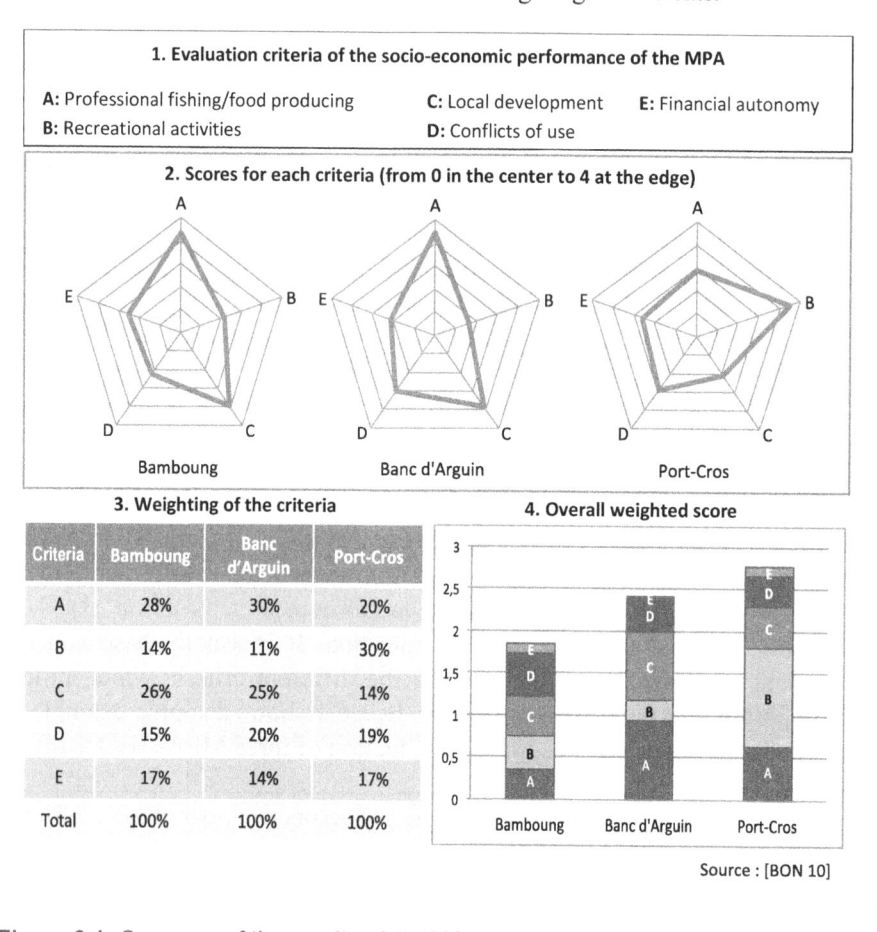

1. Evaluation criteria of the socio-economic performance of the MPA

A: Professional fishing/food producing   **C:** Local development   **E:** Financial autonomy
B: Recreational activities   **D:** Conflicts of use

2. Scores for each criteria (from 0 in the center to 4 at the edge)

Bamboung   Banc d'Arguin   Port-Cros

3. Weighting of the criteria

Criteria	Bamboung	Banc d'Arguin	Port-Cros
A	28%	30%	20%
B	14%	11%	30%
C	26%	25%	14%
D	15%	20%	19%
E	17%	14%	17%
Total	100%	100%	100%

4. Overall weighted score

Source : [BON 10]

Figure 6.1. *Summary of the results of the MCA carried out for the AMPHORE project*

Box 6.1. *An example of multi-criteria analyses applied to the socioeconomic evaluation of MPAs: the research project AMPHORE [BON 10]*[14]

14 ANR-07-BDIV-0009, 2008–2011.

The main advantages and disadvantages of each of the two approaches are summarized in Table 6.3.

	Cost-benefit analysis	Multi-criteria analysis
Main advantage	Easy to compare projects (single criterion)	Each criterion is expressed in the most appropriate unit
Main disadvantage	Difficult to monetize non-market effects	Difficult to rank the projects (heterogeneous criteria)

Table 6.3. *CBA and MCA: strengths and weaknesses*

The use of a single criterion, which underpins the CBA approach, tends to simplify the evaluation of projects. However, the establishment of this criterion assumes that each of the project's expected effects has been correctly expressed in monetary terms. This can pose serious problems, particularly when the project creates significant non-market effects. This is often the case for MPAs (for example, the value of preserved biodiversity).

Multi-criteria methods allow each criterion to be expressed in the most appropriate unit of measurement, thus avoiding the inaccuracies caused by the monetization of non-market effects. However, the down side of this is the incommensurability of qualitatively different criteria. This makes it difficult to rank the projects where the partial rankings of each criterion are not consistent (which is most often the case in practice). To overcome this difficulty, it is necessary to weight the criteria or to rank them in order of priority. This process is inevitably somewhat arbitrary. The result is a "political" problem of the composition of the panel used to determine the prioritization of the criteria. In some cases, this problem can be solved by creating a body to represent the different interests at hand ("stakeholders") and judged to be legitimate by the parties involved[15]. However, another problem, known as "Condorcet's paradox[16]", may occur since the collective preferences expressed by the panel are not necessarily transitive[17], even if the preferences of the individuals within the panel are (Box 6.2).

---

15 In the example in Box 6.1, the weightings have been carried out by a panel of experts and therefore have no "political" legitimacy".

16 Named after the French philosopher and mathematician who demonstrated this phenomenon in 1785.

17 An individual's preferences are said to be "transitive" if the fact that he prefers A to B and B to C implies that he prefers A to C.

Let the three criteria (respectively, designated by the letters A, B and C) be ranked in order of importance. There are 20 voters and the results of the vote are as follows:

Ranking Type	Number of Votes
A>B>C	7
A>C>B	1
B>A>C	1
B>C>A	5
C>A>B	4
C>B>A	2

Table 6.4. *Results of the vote illustrating "Condorcet's paradox"*

These results show that a majority of voters rank A higher than B (12 out of 20) and that a majority of voters rank B higher than C (13 out of 20). However, a majority of voters rank C higher than A (11 out of 20). In this example, the collective preference of the panel of voters is, therefore, not transitive.

Box 6.2. *Condorcet's paradox: a numerical example*

Different techniques have been developed by the creators of multi-criteria methods in order to manage the conflicts caused by the non-transitive nature of collective preferences. However, none of these techniques solves Condorcet's paradox (Arrow's impossibility theorem)[18].

6.2.2. *Methods for measuring non-market values*

The conceptual problems measuring by MCA explain why the majority of economists[19] prefer CBA as a method for project analysis. They, therefore, face the issue of expressing in monetary terms values that are not expressed as an observable market price since they are not involved in purchase and sale transactions[20]. These "non-market values" fall under two categories:

18 According to this theorem, demonstrated by the American economist Kenneth Arrow, there is no democratic rule from collective decision-making that represents individuals' preferences as a coherent social choice [ARR 51].

19 At least those who adhere to the dominant "neo-classic" view of contemporary economic thinking.

20 Accounting for market values in a CBA can also be the source of difficulties. Indeed, it frequently occurs that observable market prices do not correctly represent relative scarcities, due to distorsions caused by imperfect competition (for example, monopoly prices) or public interventions (for example, subsidies given to certain activities).

– values generated by services linked to non-market activities, such as non-commercial use of the services provided by an ecosystem (for example, subsistence or recreational fishing);

– values generated by services not linked to the use of the object in question, called non-use values[21]; these values, which are potentially very significant in the field of environmental protection, result from the fact that some people may attach importance to the existence of an object even though they make no use of it (existence value), and/or want this object to be passed on the future generations (bequest value)[22].

From the mid-20th Century, economic theory has developed a host of methods which aim to measure in monetary terms non-market values. These methods have seen numerous applications, particularly in the environmental field (for a presentation, see, for example, [BOC 07, DES 93] and [MÄL 05]). They are usually sorted into two categories:

– methods based on revealed preferences;

– methods based on stated preferences.

The field of use of the second of these is larger than for the first: while methods based on stated preferences may be applied to all types of values, methods based on revealed preferences involve only use values. These methods use a marketable aid which is associated with the use of a non-market service, in order to get the users to reveal their willingness to pay for this use (WTP).

In the field of environmental economics, two methods based on revealed preferences are commonly used: travel cost and hedonic pricing methods. The travel cost method (initially proposed to determine the use value of national parks to their visitors)[23] relies on the costs incurred by users in order to travel to the considered place of use (transport cost *stricto sensu*, potentially increased to include the opportunity cost of the time spent for travel). The hedonic pricing method relies on the price differences of tradable goods or services, whose use allows us to benefit to a greater or

---

21 Sometimes termed "passive use values", an expression which can cause confusion.
22 Between use values (market or non-market) and non-use values, we find option values, which, in a context of doubt and irreversibility, are the result of potential benefits of use that could be created by the availability of the object in the future.
23 For an example of a recent application (site on the French coast), see [BON 13].

lesser extent from the non-market goods in question (for example, the differences in house prices depending on their surroundings).

When measuring non-use values, it is necessary to call upon on a method based on stated preferences. These methods rely not on real markets but on virtual markets; created by the analyst, they are presented to the interviewees as a thought experiment in order to determine their WTP for the service in question.

In this second category, the classic method is contingent valuation. This involves creating a hypothetical scenario in which the availability of the service in question is dependent on the payment of a certain sum (for example, payment of a tax or voluntary donation for the preservation of an emblematic species). The scenario is put forward to a sample of people who are asked how much money they would be willing to pay to achieve it[24] (Box 6.3).

The contingent valuation method nowadays competes with another method based on stated preferences, called the choice experiment method. As with contingent valuation, this method relies on a survey in which a sample of people are presented with various hypothetical scenarios[25]. Each alternative is described by a number of attributes or characteristics, including a monetary value to carry out the scenario. The respondents are asked to rank the different alternatives and, based on their answers, the analyst may attempt to evaluate their willingness-to-pay for certain attributes. Compared to contingent valuation, the choice experiment method takes better account of the often multi-dimensional nature of public policy choices. Its major drawback lies in its implementation, due to the repetitiveness of the choice sets which are asked one after the other to each interviewee. Box 6.4 succinctly describes, as an example, the methodology and the results of a recent WTP evaluation for the preservation of a coral reef ecosystem, carried out on the local population using the experimental choice method[26].

---

24 One variant entails asking the people their willingness to accept compensation in return for the scenario happening. Theory and experience indicate that these two variants are not equivalent.

25 Usually including a scenario that represents the *status quo.*

26 Survey carried out as part of an IFRECOR (French initiative for coral reefs) study on the economic value of coral reefs and associated ecosystems (www.ifrecor.com).

In this study, the target population was the tourists staying at or visiting the Gulf of Morbihan. The question under consideration was their willingness to pay (WTP) for the hypothetical creation of a natural reserve. The effect of the form of payment on the responses was under particular scrutiny (two forms of payment were tested: visitor's tax and entry ticket). The implications of the reserve's potential source of funding were also analyzed. The graph in Figure 6.2 shows the responses of the interviewees (649 in total), according to the proposed form of payment.

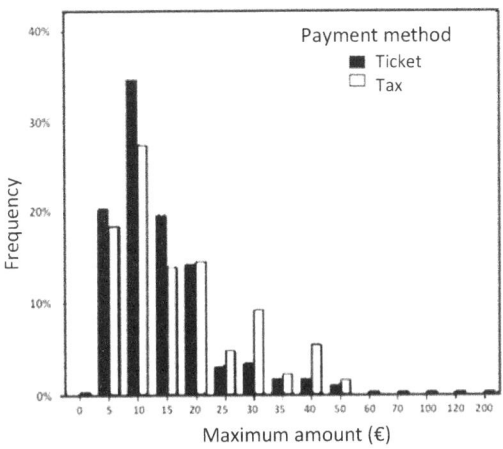

Figure 6.2. *Tourists' willingness to pay for the creation of a natural reserve in the Gulf of Morbihan ([VOL 11])*

Box 6.3. *An example of contingent evaluation applied to determining the WTP for a natural reserve project [VOL 11]*

In Box 6.4, the non-use values account for at least 30% of the total value attributed by the local population to the preservation of the ecosystem. This estimation, which, we should note, is based on conservative methodology, confirms the importance of taking into account non-use values in the socioeconomic evaluation of the benefits of coastal ecosystems[27].

---

27 Some of the respondents did not take into account the monetary attribute in the ranking of the different proposed scenarios. This question has been the subject of in-depth investigations in [MAR 14]. We will limit ourselves here to indicating that, contrary to certain *a priori* judgments, it is the respondents from the tribal Kanak populations who showed the least reluctance to including the monetary attribute in their choices.

This study had two aims: (1) to estimate the local population's WTP for the preservation of New Caledonia's coral reef ecosystem and; (2) to estimate the fraction of the WTP that relates to non-use values. For the second estimation, the main difficulty results from the fact that users (as were the populations targeted by this study) generally are struggle to distinguish use and non-use values. One solution often employed to measure non-use values entails interviewing people who are not, have not been and do not intend to use the place under study (here, people living far from New Caledonia who do not intend to ever go there could have been interviewed). However, we thus face other difficulties related to the interviewees' lack of knowledge of the subject of the survey.

The method used involved proposing conservation scenarios whose foreseeable outcomes were over different time spans: some would have effects within and others beyond the life expectancy of the interviewees. While the WTP for the first set of scenarios included use and non-use values (in indeterminate proportions), the WTPs of the second set of scenarios are solely equivalent to the non-use values. We can thus attain at an estimation of the minimum non-use values.

In total 550 people, who made up a representative sample of the population of the two areas under study (Voh-Koné-Poimbout in the Northern Province and western coastal zone in the Southern Province), were interviewed in person. These people were faced with a game of hypothetical choices in which they had to choose several (eight) times between three options that described possible scenarios for the preservation of the ecosystem, each with various attributes (quantity of fished animals, health and abundance of underwater life, preservation of the coastal landscape and the coral reef lagoon, and preservation of activity areas). At each step, two of the three scenarios required a monthly payment and a preservation of certain attributes for 20, 50 or 100 years. The third option was a *status quo* situation (no payment but degradation of the ecosystem in the long term due to anthropogenic pressures).

Table 6.5 summarizes the results of the WTP; CFP francs, the currency used in New Caledonia, have been converted into euros (1,000 CFP francs = 8.32 euros).

	Average per house (€/month)	Both zones (€/year)
Total WTP	42.05	3,025,642
Non-use value (lower limit)*	13.33	905,305

*WTP for the preservation of ecosystem attributes beyond the respondent's life expectancy

Table 6.5. *Results of the WTP for the preservation of a coral reef ecosystem [MAR 15]*

Box 6.4. *An example of the application of the experimental choice method: evaluation of WTP for the preservation of a coral reef ecosystem in New Caledonia [MAR 15]*

6.2.3. *Bioeconomic models*

Whichever method is used (CBA or MCA), the analysis of a project requires the identification and the quantification of the impact of the project on all aspects that it concerns. This can be made easier by using a model, which gives a simplified, coherent and formalized representation of the relationships that are supposed to characterize the real situation under consideration. For a project concerning an ecosystem under pressure from economic activities (fishing, tourism, etc.), this model usually has a "bioeconomic" character; it combines relationships that reflect the functioning of the ecosystem (biological aspect) and the human activities that affect it (economic aspect).

Bioeconomic modeling of MPAs has seen a rapid development over the last 20 years (Boncoeur *et al.* in [CLA 11], Chapter 8). While it is meant, in principle, to encompass all the activities that use the services of the MPA's ecosystem, in practice fishing activities have been the focus in the development of this modeling. This can be explained by the fact that, since the mid-20th Century, fisheries management has been a driving force for the development of bioeconomic modeling (for a synthesis, see, for example, [AND 10]).

Since the mid-1990s, specialists in bioeconomic modeling of fisheries have shown an increasing interest in MPAs, seen as a tool for the management of fisheries. This phenomenon has two complementary explanations: the rapid development of MPAs worldwide (see section 6.1); and the difficulties faced by the classic methods of fisheries management which succeed poorly in reducing overfishing. Faced with these difficulties, there are many advocates, especially within environmental NGOs, for the use of MPAs to improve the management of fisheries (for an in-depth analysis, see [GAR 13].

Different bioeconomic models have been created to attempt to define the role that MPAs may play in fisheries management. Although quite varied, these models are usually rather reductive: non-fishing aspects are often overlooked[28], and the MPA is usually simplified to a no-fishing area (reserve), sometimes

---

28 Some models nevertheless do integrate non-fishing aspects. See, for example, [BON 02].

adjacent to a "buffer area" in which fishing activities are permitted under certain restrictions.

The integration of the MPA into bioeconomic modeling of fisheries has required a conceptual evolution in the field. The implementation of an MPA implies, by definition, regulatory measures with a spatial dimension (for example, the banning of fishing in certain areas). However, this dimension has, in the past, been absent from bioeconomic models applied to the fishing industry[29]. The required spatialization of these models has occurred in steps.

In the first bioeconomic models of MPAs, little attention was given to space: the distribution area of a fish stock was assumed to be homogenous, and the implementation of the MPA generally implied setting aside a greater or smaller proportion of this area. The biological and economical effects of the MPA (state of the exploited resources and economic situation of fishing fleets) were considered according to the proportion of the area originally allocated to fishing that was chosen to be set aside (Box 6.5).

Despite their simplistic nature, these first-generation models enabled the study of some important phenomena, in particular those involved in spillover effects: the net exportation of exploitable biomass from a protected area to an area open to fishing. By taking into account the interaction between the biological dynamics of the resource and the economic dynamics of the fishing activity, the following points were notably able to be demonstrated:

– creating a marine reserve can be a second-best optimum when resources are overexploited by fisheries;

– it can also, in certain situations, be an insurance against the collapse of the resource (concept of "minimum level of biomass for safety");

– however, the attainment of the expected benefits of the creation of a reserve depends largely on the ability to control the fishing pressure outside the reserve area[30].

---

29 The classic models of population dynamics that constitute the biological basis of most bioeconomic models of fishery management are not spatialized.
30 This conclusion goes against a lot of the opposition, sometimes presented as fundamental, between "conventional" methods of fishery management and "MPA based" management methods.

This model is based on the red snapper fishery of the Gulf of Mexico. It simulates the effects of protecting different proportions of the fishing area on the landings, according to the level of the fishing effort (treated as an exogenous variable).

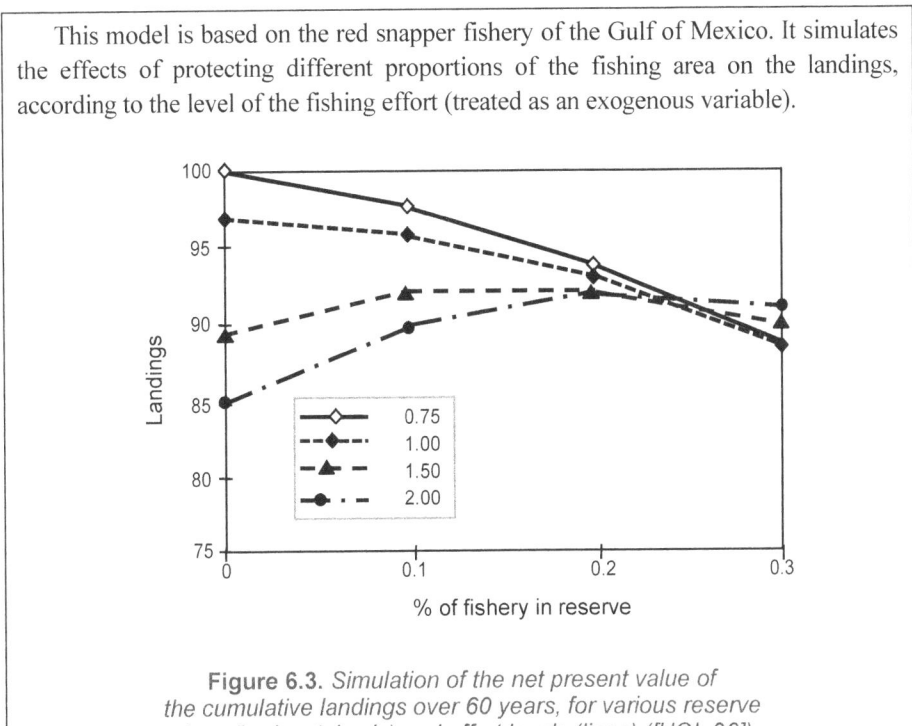

Figure 6.3. *Simulation of the net present value of the cumulative landings over 60 years, for various reserve sizes (horizontal axis) and effort levels (lines) ([HOL 96])*

Box 6.5. *An example of a spatially implicit bioeconomic model of an MPA [HOL 96]*

This first generation of MPA bioeconomic models was very quickly taken over by a second generation, named "spatially explicit" models [HOL 99, SAN 99, SAN 01]. The principle of these models was to represent "a group of metapopulations" of fish spread out in differentiated zones but interconnected. These models enabled the effect of the shape and the location of the MPAs to be studied, and not only their size. Their development required a realistic depiction of the spatial heterogeneity and the resulting processes, both on a biophysical and techno-economic level, and therefore requires a lot of data. For example, Figure 6.4 shows the spatial "grid" of a spatially explicit model[31] applied to the MPA of the Medes Islands, off the Mediterranean coast of Spain.

---

31 Developed as part of the European EMPAFISH project (FP6, SSP8-006539).

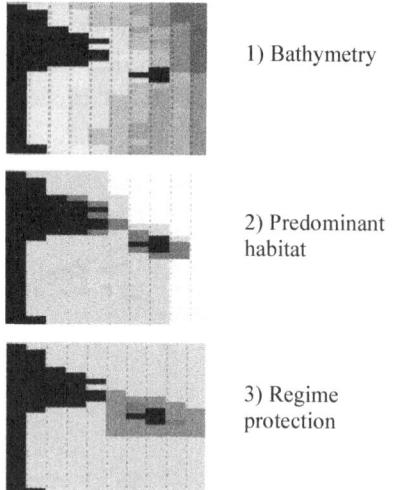

1) Bathymetry

2) Predominant habitat

3) Regime protection

Figure 6.4. *A spatially explicit model: cartography of the BEAMPA model [MAY 08]*

REMARKS ON FIGURE 6.4.– The space is represented in the form of a grid of contiguous cells, with three permanent features mapped onto the three figures. The black cells represent land areas, the gray cells represent marine areas, with different shades of gray according to the different values of the feature in question. Each cell is given a set of additional attributes which may vary over time: adult biomass and recruitment (by stock), distribution of the fishing effort (by fleet) and visitation for recreational activities.

Regarding the economic aspects, a spatially explicit model must represent the behavior patterns of the changes in the location of the fishing effort when partial or total restrictions are applied to a certain area. To this end, different modeling techniques can be implemented, particularly RUM[32] [HOL 99] models and multi-agent models[33] [SOU 06].

---

32 For Random Utility Models. These discrete choice behavioral models assume that, faced with a finite number of alternatives, an individual will choose the one which maximizes his/her utility, which itself is dependent on attributes which may be observable or non-observable (which are described by a random variable).

33 A multi-agent model is a system composed of a group of agents (for example, fishers), located in a certain environment (for example, a fishery subject to certain regulations), subject to behavioral routines and interacting according to certain relationships (competition, copying, cooperation, etc.).

6.3. Difficulties and adaptations

The implementation of the methods presented in the previous section may face major obstacles, which forces the evaluator to use alternative solutions with more or less satisfactory results. In this section, we will address the problems inherent in the measurement of the non-market values meant to maintain the MPAs. We will then describe the problems inherent in modeling the bioeconomic processes which the MPAs are supposed to influence. For each case, we will outline the main adaptations used in the attempt to resolve, or at least to reduce, these difficulties.

6.3.1. *Difficulties in measuring non-market values*

The methods that aim to measure non-market values in monetary terms have seen considerable development in the last half-century. They, however, remain complex and costly to implement, while their misuse may lead to biases that would seriously distort the evaluation.

This discussion will be restricted to the problems inherent in the application of the methods based on declared preferences, which are the only ones which enable the measurement of non-use values.

The main difficulty in the implementation of these methods is their hypothetical character: whether it involves a contingent valuation or experimental choice, the respondents are asked to make a statement about situations not from the "real world"[34], even if they may be close to it. Thus, for a survey of contingent valuation, the interviewees are normally asked to state how much they would be willing to pay for a situation that, by definition, does not exist. This type of question can cause serious misunderstandings. One of the most often encountered is the interviewees' misunderstanding of the presented scenario (particularly if it involves non-use values) or the unreal implications on their personal budget. In this last case, the interviewees may not give any consideration to the proposed sums to be paid during the evaluation, which renders the estimation of their WTP useless. If not correctly detected, this phenomenon can seriously bias the conclusions of the evaluation. Another source of bias is due to what is known as "boycotting"; some people state zero WTP, not because they do not attach any value to the proposed scenario, but because they consider that

---

34 Except for the *status quo*, which normally is one of the optional scenarios proposed.

they should not have to pay for its implementation. A slightly different difficulty arises from answers that reflect "strategic behavior" from certain respondents who underestimate or overestimate their true willingness to pay in order to manipulate the results of the survey. It has also been observed that the proposed form of payment and the means to declare one's willingness to pay could significantly influence the answers (for example, see [VOL 11])[35].

These different issues were fiercely debated following the pollution generated by the grounding of the Exxon Valdez oil tanker off the Alaskan coast in 1989. The legitimacy of the use of the contingent valuation method to evaluate the ecological damage caused by this pollution found itself at the heart of a huge legal controversy, and leading experts were called upon by both sides at the trial. The result was that in 1993 a prestigious panel (which included two Nobel prize winners in economics) was created, under the auspices of the NOAA[36], to make recommendations for the "proper use" of contingent valuation [ARR 93]. The resulting report is still today considered as the good practice guide for contingent valuation.

However, beyond the technical difficulties, the very legitimacy of attempting to translate into monetary terms certain non-market values, particularly non-use values, remains a subject of controversy (for example, see [DIA 94]). A strong basis for the attempts relates to the agents' preferences which are assumed to be those of the *homo oeconomicus* from the standard microeconomic theory. In this behavioral model, called "substantive rationality", the individual is able to coherently rank, in order of preference, all of the combinations of goods or services that he might purchase[37] and choose infallibly that which maximizes his satisfaction within his budgetary constraints. The lack of consideration given by this model to the informational and cognitive problems that real-life people face when making decisions has led to the development of alternative models, said to be of "bounded rationality" or "procedural". The individuals make choices that they judge "reasonable" given the available information and their limited

---

35 See also Figure 6.1.

36 National Oceanic and Atmospheric Administration, an American Federal Agency responsible for the study of the ocean and atmosphere.

37 Certain combinations may be ranked *ex æquo*, which reflects the fact that they bring the individual the same level of satisfaction (formally, the individual in standard microeconomic theory is supposed to have a relation of total preorder for all of the combinations of goods or services that he might acquire).

ability to process this [SIM 97]. Another criticism of the standard model is its assumptions on the very nature of preferences which involve trade-offs between alternative goals (such as to give up a certain quantity of good A in exchange for a greater quantity of good B). In certain cases, individuals may not accept such a trade-off, deeming that a part of their goals must be fulfilled whatever the cost ("lexicographic" preferences)[38]. Faced with this type of behavior, the very notion of willingness to pay for the preservation of an ecosystem service (for example) loses all meaning: the interviewee is not prepared to exchange this preservation against any quantity of money (or for whatever else).

These issues explain why non-use values are not always included in CBAs applied to the evaluation of MPAs. This results in an underestimation of the value of protecting ecosystems and a potential bias toward the development of commercial activities inside MPAs.

Some analyses try to get around the practical issues involved in the implementation of methods for estimating non-market values by using the "benefit transfer" method. This involves using previously measured values in a situation that is more or less similar (for a review, see [JOH 10]). The advantage of this method is the speed and ease with which it may be implemented (it does not require field surveys). This explains its large popularity, especially with consultancy firms. However, due to the unique nature of each site, its implementation can cause serious problems, which has prompted an expert panel created by the United States environmental protection agency to recommend extreme caution in its use [USE 09]. This caution has not always been heeded in the numerous recent attempts to measure the total economic value of certain ecosystems or the non-market benefits given by certain MPAs.

Unable to correctly measure the advantages of protecting ecosystems, some evaluations have abandoned this measure in order to concentrate on the costs of the protection. The objectives of projects are thus determined *a priori* and the aim is to decide how to attain these objectives as cheaply as possible. This approach is a characteristic of the cost-efficiency analysis (CEA), which constitutes a "watered-down" version of the CBA (Box 6.6).

---

38 In the case of lexicographic preferences, an agent prefers any amount of good A to any amount of another good B.

The Marxan model [BAL 00] involves applying a CEA approach to MPAs. It aims to determine the optimal configuration of an MPA (by minimizing the total cost of the protection that it achieves), while taking into account certain predefined protection constraints. Figure 6.5 illustrates an application of this method to an Australian MPA.

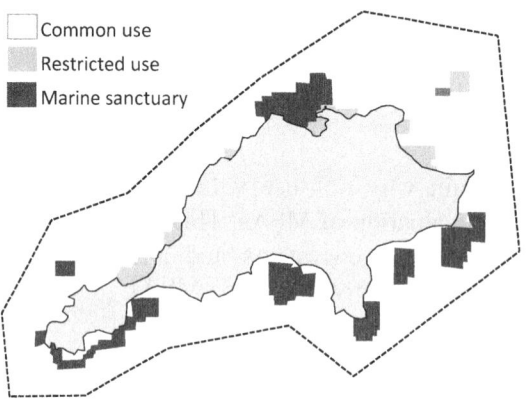

Common use
Restricted use
Marine sanctuary

Figure 6.5. *Application of the Marxan model to the zoning of the multi-use marine park of Rottnest Island, Western Australia (from [WAT 09])*

Box 6.6. *An application of CEA to determine the configuration of MPAs: the Marxan model*

The disadvantage of the CEA approach, compared to the CBA, is that it cannot weigh up the costs and benefits of the protection. However, it is also this reduced ambition that, in many circumstances, makes it more functional.

6.3.2. *Difficulties in implementing operational bioeconomic models of MPAs*

Bioeconomic modeling of MPAs only emerged recently. While it has essentially been confined to the fishing aspect of MPAs, it has nonetheless seen significant developments in the last two decades. The main cause of progress in this domain has been an increasingly accurate representation of space. Indeed, this is a determining factor in two ways for the understanding of the role played by MPAs as a tool for fishery management.

First, an MPA's impact on fishing is tightly linked to the spatial mobility of the fishers' target species. The phenomena of exportation of exploitable biomass (spillover) and dispersion of larvae from the reserve (or any zone where fishing is under any kind of restrictions, such as the banning of certain gears) are decisive for compensating the opportunity cost incurred by the fishers as a result of the constraints imposed as part of the MPA.

Second, the MPA's effect on the fishers' incomes also depends on how they adapt their effort under these constraints. On the one hand, banning fishing in certain areas forces the fishers to reallocate fishing effort in other fishing grounds (this creates additional costs and, often, conflicts with other fishers), or to entirely stop their activities. On the other hand, the spillover effects can attract fishers who previously did not use this area, thus increasing the concentration of activity on the edge of the reserve (fishing the line).

These phenomena, combined with the spatial mobility of fish and fishers, have been increasingly better accounted for, at least qualitatively, by models in the last 15 years or so. However, bioeconomic modeling is still far from being an operational tool for the evaluation of MPAs, even if it is restricted to fishing, its field of specialization. The explication for this state of affairs is empirical. It is primarily due to the lack of knowledge about the effects of reserves on the resources that are exploitable for the fishers. Studies by biologists have, until now, focused mainly on what occurs inside the fishing reserves. In this field, significant results have been obtained. With a large variability depending on the characteristics of the reserve and the mobility of the species concerned, it has been shown that biomass can significantly increase inside a reserve, due to both the increased number and size of individuals making up this biomass. The species composition of the populations inside the reserves has also been studied in detail. The understanding of the effects of the creation of a reserve on this composition has benefitted from recent developments in multi-species and ecosystem modeling ([GAR 13] Chapter 6). However, empirical evidence on the mobility between the reserve and the fishing area of species of fishery value remains a lot smaller (Goñi *et al.* in [CLA 11] Chapter 3). The quantification of *spillover* effects has made little progress[39] and, due to the more general

39 Beyond the technical difficulties that it poses, the measurement of *spillover* effects must account for the active adaptation behavior of fishers. This problem mainly concerns the frequent use of a gradient of capture (by unit of effort) from the border of the reserve. This gradient is affected by the spatial redistribution of the fishing effort, and therefore cannot be considered as a good measure of the *spillover* effect.

uncertainties about the relationships between breeding stock and recruitment, the quantification of larval dispersion's effects on fishers' exploitable resources is almost non-existent. Nevertheless, this type of information is crucial for the evaluation of the role of MPAs in relation to fishing. The lack of availability of economic data on the opportunity cost of MPAs for fishers is a complicating factor[40].

Under these conditions, the MPA performance evaluation, when it is carried out, usually relies on "dashboard" indicators not linked to a model that details the relationships between the different variables that it accounts for. This may make it difficult to interpret the evolutions that these indicators show. Indeed, it is often difficult to determine what is due to the MPA and what is the result of exogenous factors. The simultaneous observation of what occurs in an unprotected "control area" can be a palliative measure, but is rarely carried out. The fact that the initial state, before the creation of the MPA, is often not assessed does not make it any easier to interpret these indicators.

The well-known methodology to establish this dashboard of MPA performance indicators was created by the IUCN [POM 04]. It is a generic methodology, but has a great flexibility, designed to allow its application for a wide range of cases[41]. It was tested in the early 2000s on 18 MPAs across the globe and takes the form of a practical manual, containing "instructions" for the creation of each indicator (Box 6.7).

The indicators in this dashboard are rarely used to make a quick evaluation of MPA performances as in MCAs. This can be explained by the problems in implementing this type of analysis (see section 6.2.1). Their interpretation usually relies on defining threshold values, often visualized using color-coding[42].

---

40 In certain studies, this cost is simply ignored, and the qualitative indicators of a *spillover* effect from the reserve serve to demonstrate the effectiveness of the MPA on fishing.
41 In practice, however, the methodology is mainly oriented toward MPAs in developing countries.
42 See, for example, www.parc-marin-iroise.fr/Le-Parc/Objectifs/Tableau-de-bord for the Iroise marine natural park, created in France in 2007.

The IUCN manual first recommends defining the MPA's objectives. These potential objectives fall under three broad fields (biophysical, socioeconomic and governance) and are structured into two levels. For the socioeconomic field, for example, six general goals are suggested:

1. enhance or maintain food security;

2. enhance or maintain livelihoods;

3. enhance or maintain non-monetary benefits to society;

4. equitably distribute benefits from the MPA;

5. maximize the compatibility between the local culture and the MPA's management;

6. promote environmental awareness and knowledge.

Each of these general goals is broken down into a series of more operational objectives. For example, goal *1* is broken down into two objectives:

1A. satisfy the nutritional needs of coastal residents;

1B. improve the availability of locally caught seafood for the population's consumption.

For each field, the manual then suggests a series of indicators meant to inform on the achievement level of the objectives. The relationship between objectives and indicators is given in the form of a matrix: some indicators relate to several objectives, and, vice versa, some objectives can be informed by several indicators. In the socioeconomic field, the manual suggests 16 indicators, expressed in various metrics (usually, non-monetary) and with a large scope ("understanding level of human impacts on resources", "perceptions of seafood availability", "perceptions of non-market values", etc.). The collection of data for the creation of these indicators relies mainly on field surveys.

Box 6.7. *MPA performance indicators: the IUCN manual [POM 04]*

6.4. Use of socio-economic evaluation of MPAs in practice

Socio-economic evaluation of MPAs is today widely recommended in the academic literature. The methods for this evaluation have greatly progressed in recent times, even if their implementation is not always simple and often requires some compromise, as we have seen in the previous section.

A review of the international literature shows that many studies of real-life situations have concentrated on evaluating the local economic effects, which is only part of the socioeconomic evaluation of MPAs (Alban *et al.* in [CLA 11] Chapter 10). There are many factors that might explain this fact. First, this restrictive approach avoids the thorny question of non-use values. Furthermore, there is a strong social demand for this type of evaluation, in so far as MPAs' acceptability is closely linked to their ability to generate income and employment for the local population, who are usually the most directly impacted by the restrictions induced by the protection of the ecosystem. In these case studies, the impacts of tourism are studied more frequently than the MPA's effect on the fishing situation. This may be explained by the previously discussed difficulties facing the measurement of the effects MPAs on fisheries[43] (see section 6.3.2). Descriptions may often be found of various means for developing, within the MPAs, "alternative livelihoods" for fishers and their families, but the real effects of these means are rarely evaluated ([GAR 13] Chapter 10).

Overall, the actual practice of socioeconomic evaluation of MPAs remains a rather poorly known subject. These uncertainties involve not only the frequency of this practice but also for what it is used. The evaluation may indeed be used for operational purposes, as a decision-making tool, or for communication purposes, in order to support decisions that have been made (for example, to demonstrate to the public and the financiers, the benefits of protecting a marine or coastal ecosystem).

This issue particularly concerns the evaluation of the ecosystem services that aim to protect the MPAs. According to a literature review carried out by Laurans *et al.* [LAU 13], very few academic publications on the evaluation of ecosystem services show an effective implementation of this evaluation as a part of a practical process for decision-making (only 2% of the articles reviewed). Liu *et al.* [LIU 10] and Börger *et al.* [BÖR 14] show that this implementation is significantly influenced by institutional and regulatory factors and therefore greatly varies depending on the field of application under study: in the United States, the evaluation of damages to natural resources and the management of forestry and water resources represent prioritized fields of application. In France, a report from the Strategic Analysis Center (an organization overseen by the Prime Minister) presented

---

43 When the effects of the MPA on fishing are covered, this is usually done based on conventional hypotheses (see, for example, [CLE 10] or [MAN 13]).

a review on the evaluation of ecosystem services with suggestions for the establishment of reference values to help public decision-making over the protection of biodiversity, while highlighting the difficulties and limits of this exercise [CHE 09]. A recent study on the Caribbean Islands showed that the evaluation of ecosystem services could help to increase the level of awareness of the economic importance of preserving coastal ecosystems. However, very few studies (about 5% of all the studies under consideration) have played a recognized direct role in decision-making over public policy [WAI 15].

A relatively simple method to discern the practical uses of the evaluation of ecosystem services is to directly interview the people involved in decision-making. In Australia, Rogers *et al*. [ROG 13] carried out interviews with experts in non-market evaluation and with governmental agency members involved in decision-making. They then compared the results of the two surveys. It was shown that, very often, the decision-makers do not have a clear understanding of what is non-market evaluation and that, even if they use it freely *a posteriori* to justify certain decisions, they only make very limited use of it in the decision-making process. Similarly, the survey carried out on the researchers suggested that they are often overly optimistic of the true impact of non-market evaluation and poorly grasp the main factors that limit the use of this evaluation (which seem to be only loosely connected to the academic debate on the subject).

An online survey with a similar approach was carried out by Marre [MAR 14], also in Australia, using a sample of people involved in the decision-making process[44] for the management of marine and coastal areas. This survey aimed to find out their thoughts on the evaluation of marine and coastal services and what use they make of this type of evaluation. The information from a total of 88 completed questionnaires was exploited. A very large majority (93%) of the respondents claimed to understand the notion of the evaluation of ecosystem services[45], and 59% claimed to use it in their work, but only 20% used it frequently. A very large majority again (95%) deemed this exercise to be useful or essential, but almost as many (84%) emphasized its limits. The most often mentioned, among these limits, is the lack of social acceptability, followed by the practical impossibility of

---

44 With an informative, advisory or decision-making role.
45 A similar survey was carried out in parallel with a representative sample of the Australian population. It showed that 80% of respondents had no understanding of the notion [MAR 14].

accounting for ecosystems' complexity. Furthermore, the perception of the use and reliability of the evaluation of marine and coastal ecosystem services varies greatly according to the field of application. Although judged very useful and relatively reliable in the field of commercial fishing, this exercise is perceived as less useful and a lot less reliable in the field of non-use values. Finally, when interviewed on the relative importance of the different types of indicators for decision-making on coastal management projects, the majority deemed ecological indicators as more important than socioeconomic indicators[46]. These results suggest that there is still much progress to be made before socioeconomic evaluation of MPAs is trusted in practice as much as it is nowadays respected in theory.

6.5. Bibliography

[AND 10] ANDERSON L.G., SEIJO J.C., *Bioeconomics of Fisheries Management*, John Wiley & Sons, New York, 2010.

[ARR 51] ARROW K., *Social Choice and Individual Values*, Cowles Foundation, Yale University, 1951.

[ARR 93] ARROW K., SOLOW R., PORTNEY P.R. *et al.*, "Report of the NOOA panel on contingent valuation", *Federal Register*, vol. 58, no. 10, pp. 4601–4614, 1993.

[BAL 00] BALL I.R., POSSINGHAM H.P., Marxan (v 1.8.6): Marine Reserve Design Using Spatially Explicit Annealing, User Manual, available at: www.uq.edu.au/marxan, 2000.

[BOC 07] BOCKSTAEL N.B., MCCONNELL K.E., *Environmental and Resource Valuation with Revealed Preferences: A Theoretical Guide to Empirical Models*, Springer, Dordrecht, 2007.

[BON 02] BONCOEUR J., ALBAN F., GUYADER O. *et al.*, "Fish, fishers, seals and tourists: economic consequences of creating a marine reserve in a multi-species, multi-activity context", *Natural Resource Modeling*, vol. 15, no. 4, pp. 387–411, 2002.

---

46 The survey made a distinction between economic indicators *stricto sensu* and socioeconomic indicators. While the former is related to the monetary value of the ecosystem services (evaluated based on willingness to pay for non-market values), the second is related to variables such as employment, turnover, the population's participation in non-commercial activities and their perceptions on the subject of biodiversity). In the responses to the survey, these two categories of indicators were completely surpassed by fishing resources and marine biodiversity indicators (ecological indicators).

[BON 10] BONCOEUR J., NOËL J.F. (eds), Rapport de synthèse sur les indicateurs socioéconomiques, Projet AMPHORE (ANR-07-BDIV-009), UBO/UVSQ/ IMROP/CRODT, 2010.

[BON 13] BONCOEUR J. (ed.), Evaluation et suivi des effets économiques de la fréquentation des sites littoraux et insulaires protégés: application aux îles Chausey et au Mont-Saint-Michel, Projet BECO, programme LITEAU III, rapport final, UBO, AMURE/LETG GEOMER, Brest, available at: www.umr-amure.fr/pg_electro_rap.php, 2013.

[BÖR 14] BÖRGER T., BEAUMONT N.J., PENDLETON L. et al., "Incorporating ecosystem services in marine planning: the role of valuation", Marine Policy, vol. 46, pp. 161–170, 2014.

[BRO 01] BROWN, K., ADGER W., TOMPKINS E. et al., "Trade-off analysis for marine protected area management", Ecological Economics, vol. 37, no. 3, pp. 417–434, 2001.

[CGD 14] CGDD, Analyse multicritères des projets de prévention des inondations: Guide méthodologique, avaialble at: www.developpement-durable.gouv.fr, 2014.

[CHE 09] CHEVASSUS-AU-LOUIS B., SALLES J.M., PUJOL J.L., Approche écosystémique de la biodiversité et des services liés aux écosystèmes. Contribution à la décision publique, Conseil d'Analyse stratégique, La Documentation Française, Paris, 2009.

[CLA 11] CLAUDET J. (ed.), Marine Protected Areas: A Multidisciplinary Approach, Cambridge University Press, Cambridge, 2011.

[CLE 10] CLÉMENT T., GABRIÉ C., MERCIER J.R. et al., Aires Marines Protégées – Capitalisation des expériences cofinancées par le FFEM – Partie 2, Rapport 8: "Evaluation économique et calcul du taux de rendement interne des projets d'AMP", Oréade-Brèche/ FFEM, 2010.

[COS 97] COSTANZA R., D'ARGE R., DE GROOT R.S. et al., "The value of the world's ecosystem services and natural capital", Nature, vol. 387, pp. 253–260, 1997.

[DES 86] DE SILVA M., GATELY E.M., DESILVESTRE I., A bibliographical listing of coastal and marine protected areas: a global survey, Woods Hole Oceanographic Institution Technical Report WHOI-86-11, Woods Hole, MA, 1986.

[DES 93] DESAIGUES B., POINT P., Economie du patrimoine naturel – La valorisation des bénéfices de protection de l'environnement, Economica, Paris, 1993.

[DIA 94] DIAMON P.A., HAUSMAN J.A., "Contingent valuation: is some number better than no number?" Journal of Economic Perspectives, vol. 8, no. 4, pp. 45–64, 1994.

[DOD 09] DODGSON J.S., SPACKMAN M., PEARMAN A. *et al.*, Multicriteria Analysis: A Manual, Department for Communities and Local Government, London, available at: www.communities.gov.uk, 2009.

[GAR 13] GARCIA S.M., BONCOEUR J., GASCUEL D. (eds), *Les aires marines protégées et la pêche: bioécologie, socioéconomie et gouvernance*, Presses Universitaires de Perpignan, Perpignan, 2013.

[HOA 95] HOAGLAND P., KAORU Y., BROADUS J.M., A methodological review of net benefit evaluation for marine reserves, Paper no. 027, Environmental Economics Series, World Bank, 1995.

[HOL 96] HOLLAND D.S., BRAZEE R.J., "Marine reserves for fisheries management", *Marine Resource Economics*, vol. 11, pp. 157–171, 1996.

[HOL 99] HOLLAND D.S., SUTINEN J.G., "An empirical model of fleet dynamics in New England trawl fisheries", *Canadian Journal of Fisheries and Aquatic Sciences*, vol. 56, pp. 253–264, 1999.

[JOH 10] JOHNSTON R.J., ROSENBERGER R.S., "Methods, trends and controversies in contemporary benefits transfer", *Journal of Economic Surveys*, vo. 24, no. 3, pp. 479–510, 2010.

[KEL 95] KELLEHER G., BLEAKLEY C., WELLS S. (eds), *A Global Representative System of Marine Protected Areas*, 4 volumes, IUCN/The World Bank/GBRMPA, Washington DC, 1995.

[LAU 13] LAURANS Y., RANKOVIC A., MERMET L. *et al.*, "Actual use of ecosystem services valuation for decision-making: questioning a literature blindspot", *Journal of Environmental Management*, vol. 119, pp. 208–219, 2013.

[LIU 10] LIU S., COSTANZA R., FARBER S. *et al.*, "Valuing ecosystem services theory, practice, and the need for a transdisciplinary synthesis", *Annals of the New York Academy of Sciences*, vol. 1185, pp. 54–78, 2010.

[MÄL 05] MÄLER K., VINCENT J. (eds), *Handbook of Environmental Economics, Volume 2: Valuing Environmental Changes,* North-Holland, Amsterdam, 2005.

[MAN 13] MANGOS A., CLAUDOT M.A., Economic study of the impacts of marine and coastal protected areas in the Mediterranean, Plan Bleu Paper no. 13, Valbonne, 2013.

[MAR 14] MARRE J.B., Quantifying economic values of coastal and marine ecosystem services and assessing their use in decision-making: applications in New Caledonia and Australia, PhD Thesis, UBO, Brest, 2014.

[MAR 15] MARRE J.-B., BRANDER L., THÉBAUD O. *et al.*, "Non-market use and non-use values for preserving ecosystem services over time: a choice experiment application to coral reef ecosystems in New Caledonia", *Ocean & Coastal Management*, vol. 105, pp. 1–14, 2015.

[MAY 08] MAYNOU F., Results of the bio-economic and cost-benefit analysis of selected case studies, FP6, Project no. SSP8-006539 EMPAFISH, deliverable D25, April 2008, available at: www.um.es/empafish, 2008.

[PAS 11] PASCAL N., Cost-benefit analysis of community-based marine protected areas: 5 case studies in Vanuatu, Component 3E, Project 3E1 "Economics and Socio-Economics of Coral Reefs Study Report", CRISP, 2011.

[PEA 06] PEARCE D., ATKINSON G., MOURATO S., *Analyse coûts-bénéfices et environnement. Développements récents*, OCDE, Paris, 2006.

[POM 04] POMEROY R.S., PARKS J.E., WATSON L.M., *How Is Your MPA doing? A Guidebook of Natural and Social Indicators for Evaluating Marine Protected Area Management Effectiveness*, IUCN, Gland/Cambridge, 2004.

[ROG 13] ROGERS A.A., KRAGT M.E., GIBSON F.L. *et al.*, "Non-market valuation: usage and impacts in environmental policy and management in Australia", *Australian Journal of Agricultural and Resource Economics*, vol. 57, pp. 1–15, 2013.

[ROY 68] ROY B., "Classement et choix en présence de points de vue multiples (la méthode Electre)", *Revue Française d'Informatique et de Recherche Opérationnelle*, vol. 2, no. 8, pp. 57–75, 1968.

[ROY 93] ROY B., BOUISSOU D., *Aide Multicritère à la Décision: méthodes et cas*, Economica, Paris, 1993.

[SAN 99] SANCHIRICO J.N., WILEN J.E., "Bioeconomics of spatial exploitation in a patchy environment", *Journal of Environmental Economics and Management*, vol. 37, pp. 129–150, 1999.

[SAN 01] SANCHIRICO J.N., WILEN J.E., "A bioeconomic model of marine reserve creation", *Journal of Environmental Economics and Management*, vol. 42, no. 3, pp. 257–276, 2001.

[SIM 97] SIMON H., *Models of Bounded Rationality: Empirically Grounded Economic Reason*, vol. 3, MIT Press, Cambridge, MA, 1999.

[SOU 06] SOULIÉ J.C., THÉBAUD O., "Modelling fleet response in regulated fisheries: an agent-based approach", *Mathematical and Computer Modelling*, vol. 44, pp. 553–564, 2006.

[USE 09] U.S. EPA, Valuing the protection of ecological systems and services, EPA-SAB-09-012, Washington DC, 2009.

[VIL 02] VILLA F., TUNESI L., AGARDY T., "Zoning marine protected areas through spatial multiple-criteria analysis: the case of the Asinara Island National Marine Reserve of Italy", *Conservation Biology*, vol. 16, no. 2, pp. 515–526, 2002.

[VOL 11] VOLTAIRE L., NASSIRI A., BAILLY D. *et al.*, "Effet d'une taxe et d'un droit d'entrée sur les consentements à payer des touristes pour de nouvelles réserves naturelles dans le golfe du Morbihan", *Revue d'Etudes en Agriculture et Environnement*, vol. 92, no. 2, pp. 183–209, 2011.

[WAI 15] WAITE R., KUSHNER B., JUNGWIWATTANAPORN M. *et al.*, "Use of coastal economic valuation in decision-making in the Caribbean: enabling conditions and lessons learned", *Ecosystem Services*, vol. 11, pp. 45–55, 2015 (in press – corrected proofs available at: www.sciencedirect.com/science/article/pii/S2212041614000813), 2015.

[WAT 09] WATTS M.E., BALL I.R., STEWART R.R. *et al.*, "Marxan with zones: software for optimal conservation-based land- and sea-use zoning", *Environmental Modelling & Software*, vol. 24, no. 12, pp. 1513–1521, 2009.

Integrated Management of Seas and Coastal Areas in the Age of Globalization

7.1. Introduction

To broach the subject of the integrated management of seas and coastal areas raises the question of their development – sustainability being a major objective of this. This issue has been the subject of many publications, often of a rather technical and promotional nature (or the reverse). They usually take for granted that their approach is global, whereas a review of some 40 years of practice shows to what extent knowledge of "integrated management of coastal areas" has remained local, with a mainly land-based understanding of the field. Despite all the good intentions, in reality we are a far from synthesizing the different philosophies of the environmental, economic, socio-political and strategic domains; the goal being to integrate the knowledge of the wide range of disciplines in the fields of natural and human sciences.

Since coastal areas were first included in public policy, the number of concepts has multiplied; in the era of globalization, where the sea provides for all possible interconnections: human and non-human, universal and political, natural history interlinked with social histories, ebb and flow, ecology and economy, from sovereignty to world governance, sanctuary and network, enjoyment to catastrophe, etc. Oceans and coastal areas are hotspots of global phenomena and their consequences (climate change, bioinvasion, waste, pollution, piracy, migration, etc.). The answers to these

Chapter written by Yves Henocque and Bernard Kalaora.

problems are applied locally, but must be reflected on globally, thus requiring shared governance. This presupposes the coordination of state, interstate and suprastate actors, as well as cooperation with different actors in civil society. The growing awareness of the global issues, and the role of the seas and shorelines, is recent and vague. It manifests itself by the engagement of science and law, and by the ability to create effective governance in a fast-changing world.

In the following text, we propose to associate the viewpoint of integrated management expertise with that of socio-anthropology, from the angle of social engineering. Thus, we may better understand and analyze the shifting paradigms and practices which, in the name of sustainable development, underlie the integrated management of the sea and shoreline in both landscape heritage, ecological and environmental terms.

7.2. The context for integrated management practices

7.2.1. *From coastal heritage to the planet ocean*

For the first time, the seas and oceans were on the agenda at the last United Nations Conference on Sustainable Development (Rio+20, 2012). However, scientists have long been raising the alarm over the effects of climate change and other human activities on the oceans' health. The effects on this enormous space, which represents more than 99% of the livable space of our planet due to their huge depth, are such that some scientists often call this the era of "world oceanic change", rather than "world climate change". It is estimated that roughly one-third of the carbon dioxide produced by man, and 80% of the residual heat, have been absorbed by the oceans. This has probably irreversibly changed the physical and chemical characteristics, and thus the biology, of the oceans[1].

Today, the issues linked with the changes in the chemical and physical composition of the atmosphere and oceans arise from humans' behaviors.

Roughly 50% of the world's population lives in coastal areas, which represent about 10% of the Earth's surface. This leads to huge pressures on the coastal habitats and resources. In addition, most of the world's

---

1 For further information, refer to the books [MON 14a] and [MON 14b] from the "Seas and Oceans" Set.

ever-increasing population depends on oceans for food, sewage or waste disposal, energy production, or for the maritime transport which is crucial to the global economy. These people see the coast as an inspiration and a special place for recreation. Managing all of these uses, and the expectations of an ever-increasing coastal population, is a huge challenge for all countries, developed or developing.

Further offshore, the 1982 Convention on the Law of the Sea recognizes the freedom to use waters located beyond nation's economic exclusive zones. This represents more than 60% of the planet. These freedoms of use (which include fishing, navigation and underwater cable-laying) are subject to certain obligations. However, there is no coherent global system to ensure that the states and vessels that use these waters respect these commitments[2]. There are only agreements and separate organizations for vessels' discharges, hydrocarbon pollution and land-based pollution. Different organizations exist for the fishing of tuna and other species, and there are different organizations depending on the region. There is little or no coordination between these organizations. The policing of the commitments is uncertain and there are many major weaknesses in the system. These lead, for example, to persistent issues of illegal, unregulated and unreported fishing. This is estimated to represent roughly 30% of the global catch.

The lack of a global governance system means that, outside of waters under national jurisdictions, there is no integrated international mechanism to ensure the protection of vulnerable marine ecosystems from actual or potential threats. The Convention on the Law of the Sea[2] provides for environmental impact studies for activities which could cause significant or damaging changes on the marine environment. However, there is no procedure for international monitoring to ensure that these studies are carried out before the start of the activities. Such procedures would have helped to check the rapid increase in trawling, and thus to avoid the loss of rare and very vulnerable deep-water ecosystems. The Secretary General of the United Nations' attempt to launch a new deal for the oceans in June 2012 was certainly commendable. However, it did not last long due to poor dealings with the developing countries represented by the G77. In order to move toward greater coherence for "healthy oceans in a prosperous world", we must give ourselves the means for true coordination between the agencies of the United Nations. At the same time, progress must be made with

---

2 Please refer to the chapters on maritime law in this book: Chapters 1, 2 and 3.

international negotiations for further adapting the law of the seas as "common heritage for mankind".

This situation is compounded by the increasing threat from the effects of climate change. The warming of ocean waters has an influence on ecosystems' productivity and on the migration of fishing stocks. The rising sea level poses real challenges to coastal and island systems, threatening the very existence of certain small island states. The acidification of the oceans could threaten the calcification ability of corals and molluscs, and numerous planktonic species[3]. It should be noted that the questions are now asked in the context of globalization. One of the characteristics of this is the interdependence of natural and anthropic systems, according to the notion of a multipolar and transgovernmental world formed of a large number of networks (ecological, economic, social and ethical), all of which are tightly interlinked but with little local reach.

Preconditions for an ecosystem approach	Implementation and change in behavior	Attainment of environmental and societal objectives
– Coastal states commit to sustainable development of the maritime spaces under their jurisdiction (17.5) – Functional implementation of national and local coordination mechanisms (17.6) – Funds made available. The estimated total annual cost (1993–2000) to implement the programme's activities was 6 billion USD including 50 million USD for the international community based on subventions or contractual agreements (17.12) – Development of education and training for integrated management of the sea and shoreline	– Implementation of an integrated policy and a decision-making process that includes all the sectors to ensure compatibility and balance of the uses (17.6) – Development of observation, analysis and information transferral systems (17.13)	– Maintain biodiversity and productivity of habitats and marine species (17.7) – Improvement of coastal amenities, particularly housing, drinking water and treatment of waste water, refuse and industrial effluents (17.6) – Restoration of deteriorated coastal habitats (17.6)

Table 7.1. *Outline of integrated management of coastal areas as defined in Chapter 17 of the Agenda 21*

3 See Chapter 3 of [GAT 14].

In order to face these challenges, over 20 years ago, the 1992 United Nations Conference on Environment and Development, consolidated in 2002 (Johannesburg) then in 2012 (Rio), recognized the importance of implementing new forms of governance against all these changes, for which man and his activities were most often responsible. The main principles and methods of applying these new forms of governance were defined for water (integrated management of water resources) and coastal areas (integrated management of coastal areas). This was carried out with close ties to the Convention on Biodiversity (1992) and the Reykjavik Declaration on responsible fishing (2001).

All of this did not come from one single conference but from much prior work. This is brilliantly summarized in the Brundtland report (1987), a text that truly founded the concept of sustainable development.

7.2.2. A forward-thinking international impetus

Unlike the concept of linear development, of which the gross domestic product (GDP) is a perfect example[4], sustainable development introduces movement by arranging itself over two axes: one vertical (present time) and the other horizontal (the future). It can thus be interpreted as a three-dimensional system, the past (planetary ecosystem, inheritor of more than 4.5 billion years of history), the present (the relationship between the ecosystem, economic and social) and the future – all of which have needs that cannot be compromised.

This concept – while connecting economics, ethics and history – remains embedded in a context of liberal tradition. This is because it implies that the market should be recognized as a regulating system. It also diminishes the state's role compared to that of civil society, while underscoring the moral principle everyone's responsibility for future generations.

The Brundtland report was a real turning point, which would be fully realized at the Rio Conference in 1992. It initiated a shift in the collective aspiration to think about and to negotiate in new ways the relationship between human societies and nature, and thus with space and time. The ambition of this was to create a new global dynamic, encouraging people to

---

4 See the excellent website "Redefining Progress – The Nature of Economics", http://rprogress.org/sustainability_indicators/genuine_progress_indicator.htm.

believe and think that nothing would be the same again. In this sense, the Brundtland event was pivotal: it instituted globalization by the concept of sustainable development. It preceded the other milestone moment which would reinforce this globalization movement: the fall of the Berlin Wall in 1989. This would come to be a symbol of the opening and generalized increase in exchanges. The Brundtland report is based on the value ideas taken from every expert in democracy, rights of humans and nature, markets, quality of life, food safety, and personal and patrimonial property [KAL 99].

Similarly, coastal areas began to be considered in public policy in the 1970s. This was with a view to spatial planning and conservation of coastal and lake landscapes, in the face of growing urbanization. For example, in France, the acquisition of natural spaces was selected as the favored means for habitat and species conservation, after the creation of the Coastal and Lake Shore Conservation Authority in 1975 [KAL 10]. Before this, the United States more successfully passed their Coastal Management Act in 1972[5]. In either case, the subject of the sea is approached indirectly, with a land-based perspective, and thus more or less reduced to where it touches the land.

This phase was followed, in the 1980s, by a new legal system for the high seas[6]. It should be noted that, before this, the oceans were divided into two types of space. The vast majority had totally free access, especially for fishing. The other type was a thin strip, usually of three nautical miles from the coast. Here, all activities, and fishing in particular, were regulated by the coastal states. However, from the middle of the 20th Century, the situation was becoming increasingly chaotic because of the increase in ships' power and fishing capabilities. This leads states to individually stipulate their own limits for ocean "territorial rights". It has thus been a long negotiation process, full of successes and failures, starting in 1958 and finishing in 1982 with the United Nations Convention on the Law of the Sea, which was ratified 10 years later to then enter into force.

The United Nations Convention on the Law of the Sea is, therefore, a difficult compromise. It is based on the new idea that common property cannot exist without legal protection (property or the "common heritage of mankind"). This international convention set legal benchmarks for a global

---

5 A special edition of the *Coastal Management Journal* was published in 2012 to resume the 40 years that the law has been applied (40 years of the CZMA: Impacts and Innovations).
6 For more information, see Chapter 1.

policy on maritime space. It ordered countries with a coastline to define their economic exclusive zone (EEC) and continental shelf. This could lead to extension requests for the latter.

Between 1990 and 2002, a dual evolution on the perception of coastal and maritime issues gradually appeared. This came through the concept of "sustainable development", then through the concept of the necessary "economic growth". The importance of working with the private sector was emphasized, as it was at the World Summit of Johannesburg in 2002. This later evolved into the concepts of "green growth" and "blue growth". Structuring of the maritime domain, while already strongly governed by the United Nations Convention on the Law of the Sea, progressed to a second level when a whole chapter on seas and oceans was included in the Agenda 21, Chapter 17[7]. This now extends to all of the EEZ and the high seas, in continuity with the land–sea interface and its river basin, with a broad view of the social and ecological system.

This international scientific and political engagement has meant that maritime issues have been increasingly under consideration. In Europe, this has put the issue back on the agenda from 1999, with a demonstration programme on integrated management of coastal areas. This resulted in a recommendation being made in 2002 which led to a draft integrated maritime policy in 2006, following on from several other developed, emerging and developing countries throughout the world.

7.2.3. How do coastal and maritime areas lend themselves to the globalization game?

"Globalization has led to an increase in the force maritime challenges, in terms of flows as for resources. The increasing economic, diplomatic and ecological importance of maritime spaces to globalization renders the sea, more than ever, a political challenge, thanks to which a state can shine and assert its power on the international stage". These are the words of a recent report on the maritime orientation of the French Senate [SÉN 12].

---

7 It should be noted that this Chapter 17 contains no less than seven major programmes including (1) integrated management and sustainable development of marine and coastal areas, as well as EEZ; (2) protection of the marine environment; (3) the sustainable use and conservation of living marine resources in the high seas.

In this respect, maritime transport is particularly a prime example[8]. The network of maritime routes connects trading posts or ports, nodal points of these international exchange networks. Certain lines are dedicated, for example, from an oil terminal or a nickel mine to a specialized terminal near to processing plants for these raw materials. Others are veritable highways (the Mediterranean, Channel, North Sea, etc.) on which uninterrupted lines of ships sail, linking areas of production to areas of consumption. Certain nodal points are important ports (Shanghai, Singapore, Rotterdam, etc.) which are more or less multipurpose. They serve both the hinterland, via waterways or over-land, as well as other secondary ports. This constitutes a global living network, constantly evolving according to new areas of exploitation or settlement. We are more than ever in a "maritime" century. This is certainly backed up by a long history (the "silk road" and the age of discovery), but is now governed by an international legal framework and forms of governance. These are expected to develop in national (EEZ) and international (common heritage of mankind) waters.

7.2.4. The third forgotten path: common pool resources[9]

This last concept of "common heritage of mankind", highlighted by the phenomenon of globalization, merits a little attention. It involves common places (common pool resources) without borders and not dependant on the direct sovereignty of states. This is the case for the outer atmosphere and maritime spaces (including ocean floors). It also applies to the Internet, or other kinds of space "without borders", like finance, media, NGOs or, more generally, knowledge. These spaces have in common, like maritime transport, a non-hierarchic organization around a group of networks. They connect nodal points made up by different areas of settlement, exploitation, transformation of resources or immaterial data, concentration of knowledge, or policy centres. Freedom of circulation, interconnection, continuity,

---

8 Emmanuel Desclèves, "La mer, vecteur et enjeu du future", personal communication, 2013.

9 "Common pool resources" may have two different meanings: the paradigm of "common heritage of mankind" has a universal value of common responsibility which, here, is applied to the ocean. By definition, "common pool resources" belong to the collectivity in the form of resources (e.g. water, forests and fisheries) managed by a community according to a group of rules defined with and applied by the community. The study of the methods of managing these common pool resources is at the heart of Elinor Ostrom's work (2009 Nobel prize in economics).

fluidity, reconfiguration, plasticity, circumvention, capillarity, diffusion or concentration: these are all characteristics that bring these networks closer to maritime practices established since time immemorial.

In terms of governance, it is clear that the current practices are not adapted to the management of these common pools, heterogeneous, fluid, continuous and probabilistic. The recommendations of the Rio Conference (1992) on integrated management did not apply only to the local level (as practiced until now). It is a question of associating scales from local to global levels. The practice of collective action and self-organization should be prioritized. These are recommended by Elinor Ostrom [OST 10] as "a third path of action for human societies, between privatization and state action".

Elinor Ostrom has endeavored to show that, for a long time and almost everywhere in the world, communities have been and are able to manage, in an economically optimal way, common assets, using "institutional arrangements". Alongside management via individual property laws or by the state (public good, not to be confused with common good), there can exist a third effective institutional framework. Herein, communities collectively manage common assets. Ms. Ostrom has thus shown that these institutional arrangements have enabled the collective management of numerous ecosystems without leading to their collapse. For this to work, it is crucial that there is a functional interface between the social psychology (requiring trust in each other) and the operation of institutions. In this system, common space and resources are used but not appropriated. Their use assumes the implementation of distributive justice so that no-one feels aggrieved. The profits, as well as the costs, must be shared fairly. Finally, the costs of environmental damages linked to the unintentional effects of human activities must be borne by the entire collective.

When transposed to a global scale, as in the case of oceans, the use of common heritage must be for the benefit of all of humanity, beyond selfish state or private interests. All profits from resources at stake must be fairly redistributed (in financial, technological and scientific terms), including to developing states. This involves recognizing the interdependence of ecosystems and the uses made of them. In this respect, the concept of common heritage of mankind has much in common with the ecosystem approach. This aims to shed itself of any form of fragmented management which is limited to managing one single space.

The building of a shared world requires us to reconsider our modern world-view, which is based on dualisms between subject and object, science and politics, experts and laypeople, facts and values, nature and society, equality and justice, modern and traditional, etc. In this respect, knowledge should not be disconnected from man's plans and hopes to learn how to live together, to achieve personal fulfillment and to fulfill the moral development of communities. Knowledge should not only be for the purpose of knowing more, and should not be disconnected from society's moral goals and choices. Environmental and maritime policy should aim to join the fulfillment and development of members of human communities with achieving a good environmental and ecological status. To achieve these goals, we must leave open certain avenues, to encourage creativity and imagination, and trust man's ability to adapt with success to the changes and uncertainties of an increasingly complex and connected world. The ideal of conservation would have no meaning without an ambition for individual's democracy and autonomy against institutional hierarchies and barriers, which are the biggest obstacle to living together and creative development (as defined by Bergson and Darwin). It means encouraging polycentric modes of governance, which integrate different levels – from the individual to the global level – structured around physical locations.

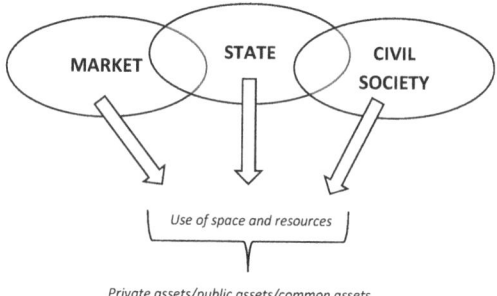

Figure 7.1. *The players in governance: the economy assisting social relations or social relations submitted to the economic system*

We will finish putting into context the integrated management practices by insisting on the importance of reducing the gap between ecology and the democratic process. This gap is always widened in the name of "economic efficiency", based only on scarcity value. This leads us to forget that which is abundant (air, water, landscapes, biodiversity, trust, etc.). In order to make these abundances more visible, it is not necessary to give them a scarcity value, but instead to give them value in the eyes of the communities that benefit from them. According to this approach, the governance, or the interaction process between the three major components – commonly termed "market", "state" and "civil society" – must be modeled above all as an issue of democracy. The devices set up and the processes implemented must "create" democracy.

Governance is built around the hard work of experts who, as part of work groups for the large international meetings (such as at Rio), have brought forth a new semantic, a semantic touching governance as functioning around four precepts:

– relegitimization and modernization of state action. This can be achieved by making its institutions more flexible and better adapted to respond to crisis situations and uncertainty;

– implementing forms of coordination, and institutional and non-institutional agreements. These should be characterized by transversal, non-hierarchical relationships. Deliberative and participative approach, renewed principle of authority (negotiation, charter, partnership, contract, international convention, instruments for mediation and subsidiarity);

– transition to flexible forms of rationality, pragmatic, reflective, which help to support the representation of individual and collective players (e.g. "sustainability" science in the scientific field);

– increased handover of decision-making power from the state to civil society.

Thus, governance refers to the unwritten institutional arrangements between the major types of player: public authorities, private players and civil society (including associations and NGOs). Governance is an organizational force which drives development toward the strategic, coherent and congruent goals of sustainability.

7.3. The ecosystem approach: dynamic interactions between societies and ecosystems

Integrated management of coastal areas and the sea used to be limited strictly to the land–sea interface. This was the interpretation made from integrated management of coastal zones (IMCZs). Nowadays, it is closely associated with the implementation of the ecosystem approach, whose principles are central to the 1992 Biodiversity Convention.

The Millennium Ecosystem Assessment [MEA 05] offered a description of the interactions between society and nature, using the concept of ecosystem services (Figure 7.2). However, it is very schematic on the "well-being" dimension and on the interactions between the four categories of capitals (physical, natural, human and social), sources of human development [LEV 07].

The main challenge related to the ecosystem approach is found in the social system, and its complex interactions with it. Holling [HOL 86] showed that ecological surprises are more likely to occur where large-scale systems become highly interconnected. This is also the case for social systems, which become more prone to instability the bigger and more interconnected they become. Under these conditions, it is not surprising that, when two unstable and highly interconnected systems meet, as integrated management addresses it, their overall stability is very uncertain. Since time immemorial, in human communities, the players in these systems have been able to coordinate their actions because they share common values.

1) The objectives of management of land, water and living resources are a matter of societal choices.

2) Management should be decentralized to the lowest appropriate level.

3) Ecosystem managers should consider the effects (actual or potential) of their activities on adjacent and other ecosystems.

4) Recognizing potential gains from management, there is usually a need to understand and manage the ecosystem in an economic context. Any such ecosystem-management programme should: reduce those market distortions that adversely affect biological diversity; align incentives to promote biodiversity conservation and sustainable use; internalize costs and benefits in the given ecosystem to the extent feasible.

5) Conservation of ecosystem structure and functioning, in order to maintain ecosystem services, should be a priority target of the ecosystem approach.

6) Ecosystems must be managed within the limits of their functioning.

7) The ecosystem approach should be undertaken at the appropriate spatial and temporal scales.

8) Recognizing the varying temporal scales and lag-effects that characterize ecosystem processes, objectives for ecosystem management should be set for the long term.

9) Management must recognize the change is inevitable.

10) The ecosystem approach should seek the appropriate balance between, and integration of, conservation and use of biological diversity.

11) The ecosystem approach should consider all forms of relevant information, including scientific and indigenous and local knowledge, innovations and practices.

12) The ecosystem approach should involve all relevant sectors of society and scientific disciplines.

Box 7.1. *The 12 principles of the ecosystem approach from the Biodiversity Convention (1992)*

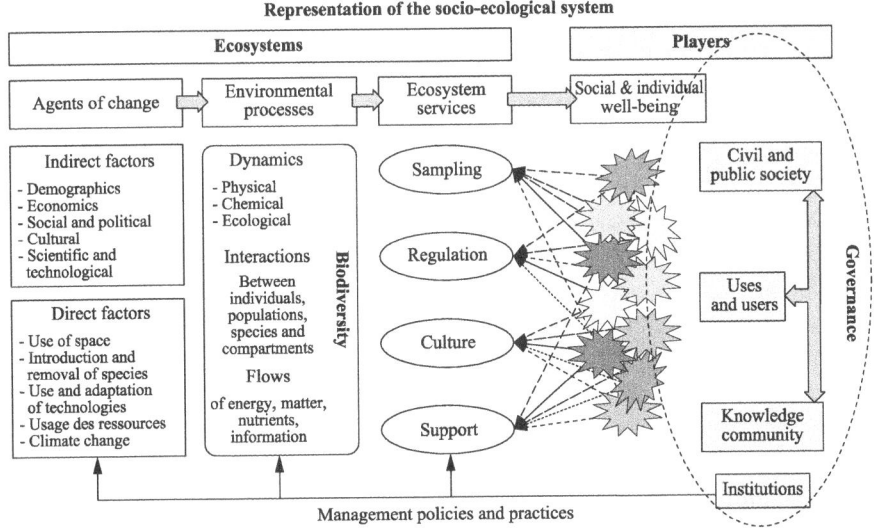

Figure 7.2. *Representation of the socio-ecological system*

In this sense, the sea and coastal areas face an increasing need for an integrated policy, in the adaptive and pragmatic meaning of the term. The players must be more in position to affirm their experience, to play their part, to use their power and responsibility. At the same time, public institutions should help to make the expression of this possible. Under these circumstances, the coastal areas and the sea may be seen as catalysts of exchange and values, raising the bar for justice and democratic requirements. They bring new life to the sector, by increasing the base to players until now ignored. This creates a dynamic where science, ethics and engagement represent major resources in the face of the uncertainties and complexities of the systems.

Therefore, a certain number of constants emerge from the dynamics of the interactions between societies and nature. It is essential to keep these in our memories when carrying out actions in the field:

– In general, a specific sector of society will be the main beneficiary of these activities. These activities will cause a great deal of stress to the ecosystem, the cost of which will have to be borne by other groups or by the whole of the society (the concept of "share the costs, privatize the benefits"). Since there are many interacting users, the goal is to get them to minimize their ecological footprint. It is also to get their activities to contribute to the tracking and/or the restoration of the ecosystem.

– The key to better sharing out the benefits is in understanding what actually directs society's dynamics in the use of the ecosystems on which it depends.

– Management and sociology researchers have shown the dynamic nature of groups of people involved in problems, and the crucial role played by the creation of social networks for the resolution of conflicts. A problem of ecosystem degradation will first be noticed by some informed people or "whistleblowers" [CHA 99]. These are often scientists or directly affected people. They will potentially try to stop or reverse the degradation, using social networks to share their cause with the largest possible number of people. If their message is clear and convincing, and if their social networks are sufficiently large, they have every chance to mobilize action. However, other interest groups might mobilize as a reaction. If the social links between the two groups are too weak, or do not exist, the polarization of ideas could stick for a long time. Only links between groups, and to higher hierarchical levels, can help in finding a solution which might lead to institutionalization.

– Certain political and social processes can lead to a compromise. In fact, it has often been shown that compromise adds little, or can, in the end, be harmful for the entire community.

7.4. Multi-dimensionality and expertise

In this process, the integrated management expert participates in aligning the resource-population-environment-development system. Starting with pre-existing circumstances of management, he must help the community in its effort to shift the environment or the resource toward a sustainable system which can be handed down to future generations. The expert is assigned to pragmatically define a coherent management structure for the marine and coastal environments. This must bring together public policies, and make its different uses compatible. In other terms, he must create social and cultural frameworks to encourage mediation and agreement. His intervention lies at the meeting point of several elements: the land – its ecological, economic and socio-political context – and social aspirations. Before any other more specialized intervention, this is, above all, a practice that we could label as "eco-sociosystemic and anthropological engineering" [KAL 99], wherein the expert's language could be classed as "neutral speak". The terms that he uses have a weak emotional and sentimental value, but on the other hand, a strong instrumental and functional value. In this way, a reference framework and observation tools can be developed, allowing the same subjects to be grasped, despite the sociocultural and political differences.

In the face of the complexity and interconnectivity of the system, uncertainty and adventure are the basic ingredients for building knowledge on the way to expertise. The first rule is to accept other knowledge, as well as to compromise with them, or to cooperate. Thus, according to the setup, the specialist, whoever he may be, can become an ecologist, biologist, planner, legal expert and vice versa. Secondly, he must compromise with the land, of which his knowledge is inevitably flawed due to a lack of time. This is true even if he has access to a large quantity of notes and information (which are not always useable). All the attributes of an adventure are displayed: the importance of improvisation, rough understanding of situations, familiarity with the players, constant adaptation to knowledge and practices.

The expert in integrated management must, therefore, accept doubt and ignorance. He cannot wait to intervene to take control of the situation, since he is working in an uncertain world, with vague objectives which he hopes will become clear in the course of action. His success depends in particular on his ability to transform these uncertainties into opportunities for action. In terms of sea and coastal integrated management, there is no model to shoehorn the behaviors of the social actors and the local populations into a theoretical framework. This requires constant learning, an appetite for challenges and risk, and curiosity.

1. The language dimension of expertise

– Create conditions for listening to facilitate exchanges/listen in order to integrate.

– Neutral speak: find subjects that make sense to all.

– To say is to do: performative nature of discussion.

To say is to institute an action, to constitute for the different actors practical possibilities.

Categories of language which carry with them something powerful, indeed; via these categories, a political reality is created.

– From the enunciators' role: the creators and catalysts of ideas.

Sustainable development, a new formula for power which is based not anymore on power devices but on the idea of contractualization and management for the common good of land.

– The ultimate trick is to justify, without announcing it, the superiority of democracy.

– Democracy is no longer the essence of the political but the exercising of a kind of reflexive democracy.

2. Profile of a coastal areas manager

– Leadership ability.

– Strategic analysis and political process abilities:

   - to manage people and institutions;

   - strategic analysis as a tool for social marketing in order to change behaviors.

– A strategist with knowledge and experience of:

   - conflict resolution;

   - management of group processes;

   - administration of complex programmes and institutions;

   - creation and administration of transdisciplinary research programmes;

   - creation and administration of public education and participation programmes;

- programme evaluation;

- summarizing, interpreting and displaying groups of complex information.

3. On the participation of actors and the public

Why involve these parties?

– help the adoption of the programmes;

– give the necessary information (which is often not available) on the resources and their uses;

– help planners to understand the direct and indirect causes of management problems;

– help to quickly test in the field the feasibility of management measures;

– from the start, help to resolve conflicts in management modes.

The preconditions of success:

– making clear what is expected and in what time-frame;

– giving sufficient time for relevant events;

– making the programmes understandable and adapted;

– clearly separating facts from policy.

Actions to encourage meaningful participation:

– clearly identify legitimate roles;

– facilitate "equality" in negotiations;

– start with the needs of the actors;

– learn from and with the participants;

– first gain trust and make them feel comfortable;

– facilitate the creation of organized groups and their leaders.

4. Adaptive management

Learning by doing

Unlike conventional planning which demands:

– a large amount of information;

– low uncertainty;

– few deciders.

In adaptive management:

– we no longer wait 10 years for information before carrying out action;

– the programmes/plans are considered as experiments;

– carrying out the programmes/plans creates opportunities to test and improve the scientific elements of support for the action;

These opportunities are integral parts of the planning system that uses information from the carrying out of the programme/plan:

– a programme/plan which learns little will be quickly overcome by uncertainties;

– a programme/plan which learns well can last, despite low levels of initial knowledge.

5. Anthropology and sustainable development

Research on the cognitive aspects of knowledge on nature – the "wild" thought – is part of the sustainable development approach.

In the Rio Conference, several lines are given by indigenous people.

Indigenous knowledge becomes an expertise tool for management.

Biologists become ethnologists.

Indigenous people become experts; they have means for sustainable management.

The expert becomes localized, the indigenous person becomes globalized.

Box 7.2. *The vade-mecum of the expert in integrated management (from [KAL 99])*

7.5. Linkage of scales and concepts

To put it simply, through international and local expertise, the evolution of management systems will be directed by existing forms of governance. This evolution will tend toward an overall consistency of policies, whether they be sectorial, spatial planning or on transversal topics, such as research and biodiversity strategies. It applies to the land–sea interface and continues out to sea according to the requirements and the spaces under consideration. Moving from land to sea (Figure 7.3), the issue of spatial planning becomes one of strategic planning of maritime space. This is defined by the European integrated maritime policy and its environmental pillar, the framework directive "Strategy on the Marine Environment". According to international acknowledgment, these maritime areas are made up of marine ecoregions [SPA 07], which are themselves made up of large marine ecosystems[10] [SHE 10]. These are the subject of cross-border projects, particularly in the context of the dedicated program, the Global Environment Facility (GEF). Integrated management of coastal areas and the sea is, therefore, not a rigid structure, but dynamic and multi-scaled. It is based on adaptation to change, and supported by forms of governance which prioritize collective learning. This represents a shift from knowing toward a process of learning, in a world full of uncertainty [CLA 03].

Here, the devil is not so much in the details, but in the dynamics in play. Whatever the goal, it should always be put in context (time-space and politico–socio-economic). Its implementation should always be integrated with means for coordinating policies and instruments for the management of the river and marine basin, passing through the land–sea coastal interface.

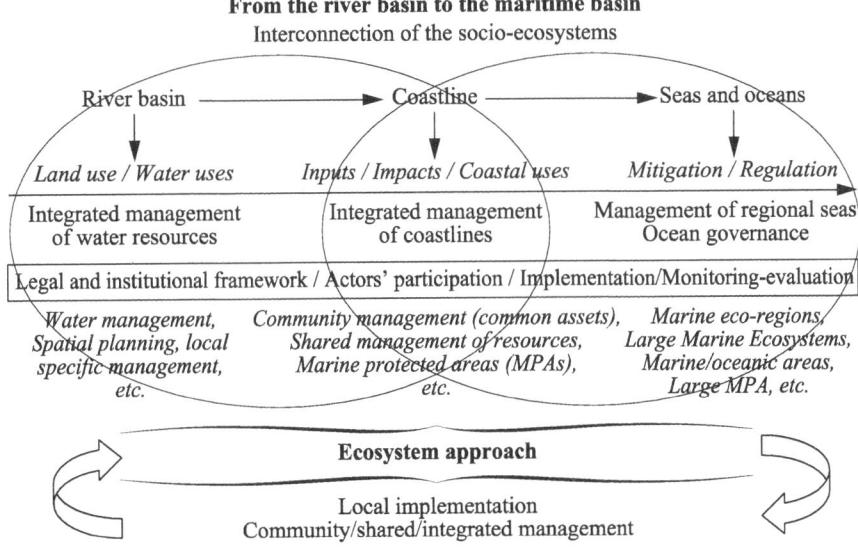

From the river basin to the maritime basin
Interconnection of the socio-ecosystems

Figure 7.3. *The operational translation of the land–sea continuum*

---

10 See the dedicated website www.lmc.noaa.gov.

7.6. Where do we stand on integrated management of the sea and coastal areas?

As we have seen, the concepts and perspectives have evolved significantly. This is reflected, in particular, by the creation of maritime strategies, both regional and national. This has occurred alongside the international debate on the necessary evolution of international sea law, to achieve a global governance of the oceans.

The ecosystem approach for the integrated management of the sea and coastal areas is now considered to be an essential tool for implementing maritime strategies. This represents an "updated governance" for the sea and coastal areas, as is on the agenda in Europe and France, as well as in many other countries and regions throughout the world.

The concept of maritime spatial planning has emerged from the different instruments for implementing these strategies. As its name suggests, it should be thought of as a concerted organizational tool for maritime activities, according to the three main elements that make up the marine environment (floor, water column and surface). It is thus an additional tool for governance and integrated management of maritime space to ensure their sustainable development (and not only their protection). As mentioned before, its implementation raises the delicate question of the inclusion of the land–sea continuum in management plans. This requires combining management tools, which were usually devised on their own, without a spatial or temporal context.

7.6.1. *Climate change, destitution and the increased vulnerability of ecosystems*

The links between climate change and the effects on coasts and marine resources need no further proof. Numerous trials to reduce and adapt to the effects of climate change are being carried out across the world. These include policy, institutional and operational trials. However, the funding commitments, made in Copenhagen, to poor countries by the rich countries have had little effect thus far. It appears that they are entirely insufficient, in light of the enormous needs to adapt to the catastrophic conditions that could result from droughts in sub-Saharan Africa, or floods in South or East Asia.

How can integrated management help in this effort to adapt? The adaptation strategies must become an integral part of the management plans adopted at a local level. These local management plans must then be absorbed at higher levels as part of national or regional strategies (e.g. regional sea conventions). Vulnerability maps and tools (such as strategic planning of maritime space, from the coastline to the high sea) are essential to defining the adaptation measures at the appropriate levels. Monitoring of water quality and habitats, such as coral reefs, contributes to the knowledge of the changes occurring. Knowledge can also be gained on the interconnections between uses and resources that underlie these changes.

7.6.2. *Persistent poverty and inequality in many parts of the world*

Whatever actions are taken, reducing the poverty and inequality in many parts of the world will not be as easy as we might think. Although in the last 10–15 years, a few nations from East Asia and South America have become emerging countries, poverty remains a major issue. It also exists in developed countries, in the United States or in Europe, particularly in the context of ever-increasing immigration.

Admittedly, one of the most recent UN reports (2010), on the achievement of the millennium development goals (MDGs), sends positive signals on the reduction of world poverty[11]. However, in time, inequalities will increase, particularly in several countries in sub-Saharan Africa and Asia, where the rich are getting richer, but a large fraction of the population remains in total poverty. The "middle" class is often too small to reduce the gap between the very rich and the very poor. These increasing inequalities contribute to increased violence, especially with the disabused and unemployed youth. It also adds to corruption in the wealthiest groups.

How can the actions of integrated management of the sea and coastal areas help? Most of the poorer classes in non-industrialized countries are dependent on natural resources for their subsistence. Improving the management of these resources could lead to changes. In particular, the local populations may become more able to take charge of their own affairs, with the local authorities. However, these more or less developed local initiatives

---

11 First objective of the MDG: halve, between 1990 and 2015, the proportion of the population whose income is less than 1 dollar a day.

are limited if they are not integrated into national policies. This would allow them to network and to gradually increase in size. The integration of environmental policies into sectorial policies is fundamental to the sustainable development of the sea and coastal areas.

7.6.3. *Increasing threat of insecurity*

Poverty and inequalities, power games and the easy access to weapons have led to an increasing lack of security in many parts of the world. Coastlines and oceans are particularly at risk due to the ease of movement that they provide. Acts of piracy, especially of the coast of Somalia, call into question the free movement of merchant ships, and fishing activities in the entire area.

How could integrated management of the sea and coastal areas aid in such a challenge? While this complex question concerns mechanisms of global governance, it just as directly involves the coastal populations of the region. Integrated management initiatives can help to mobilize these people, especially the youth, helping them to express themselves, and to become aware that they have a role to play in breaking the vicious circle of poverty.

7.6.4. *Impacts of the global financial crisis*

In the current situation, bilateral and multilateral donors are, inevitably, more careful with the investments that they undertake. Investments in the field of sea and coastline management are in competition with other priority areas such as agriculture, infrastructure, energy, health and education. Unless a programme can combine environmental protection with the reduction of poverty, as well as the growth and potential development of infrastructure, it is difficult to keep the sea and coastline high on the agenda. Every regional sea has seen the introduction of this type of strategy, such as the GEF partnerships, or international initiatives, such as the Union for the Mediterranean (UM). The NGO community is also very dependent on donors. It has not been spared as well and has had many severe cuts in programmes and staff. On the national and local levels, some countries have suffered more than others from the reduction or disappearance of markets, the lack of available credit and the reduction in tourism. The decrease in revenues has led to a decrease in funds available for infrastructure, water, energy and social programmes.

How can the ecosystem approach for integrated management of the sea and coastal areas alleviate this situation? First, the fact that its basic principles and framework for action were negotiated for and adopted by international agreements gives it credibility in the donor's eyes. Using this strategy, specific actions can be instituted, such as the creation of regional networks of marine protected areas (MPAs). These can be accompanied by strategic, negotiated planning of the maritime space that makes up the large marine ecosystems. These instruments can contribute to a more balanced growth of coastal and maritime activities.

7.6.5. *Unfair trade of marine products, the absence of capabilities and effective structures for the redistribution of benefits*

For poor countries that lack capital, trade is the solution. It does not matter to them if this trade is linked to aid measures, or if the commercial agreement profits another country or a private entity. Some government officials see this as an opportunity for political and economic progress, while others see opportunities for corruption. Trade should be a source of revenues and a means to reduce national debt and the country's dependence on loans. However, trade often simply makes some people rich to the detriment of others. It may be a nation or a group of nations which benefits, by passing unfair agreements for fishery exploitation [CUR 08]. It may be an oil company, operating with an agreement whose terms are not necessarily fair. Everywhere, who prevails depends on the balance of power, but there are a certain number of internal causes for the profitability of trade: the absence of well-established policies and institutions, the lack of ability to negotiate a fair deal, the lack of abilities and means for accounting and monitoring, state corruption and the absence of an established structure for redistribution of profits. The losers are at all levels: national treasury, civil society, the poor and all those left to fend for themselves.

In such a situation, what can the ecosystem approach for integrated management of the sea and coastal areas contribute? According to the 12 principles of the Biodiversity Convention, the ecosystem approach contributes to the development of a transparent set of practices and operating rules for the exploitation of resources and the sharing of profits. The forms of governance and integrated management which govern these practices improve the ability to implement a coherent political and legislative framework to manage these resources (and thus the ecosystem services). These include fishing, energy and mineral resources, as well as the resources necessary to develop tourism, infrastructure and urbanization. However, to speak of the transparency of the system leads us to address the question of governance.

7.7. Toward new challenges and new forms of governance

Questions of governance concern developed, as well as developing, countries. However, the consequence on developing countries can be even more crippling when there are no safeguard mechanisms.

States must address the question of governance in order to handle the growing complexity of the societies that they are meant to govern. They must explore what will be the society's forms of participations – for both the private sector and civil society – in the decision-making process. "Overall, governance does not include only intergovernmental relationships, but also non-governmental organizations, citizens' movements, multinational corporations, and the global capital market. Each of these interacts with the global media networks, whose influence is increasing" (definition of the Commission on Global Governance, 1995). Whatever the scale of the intervention, "good governance" requires much institutional capacity, in which governments pay a central but non-exclusive role.

Operationally, this governance may be applied to four séts of activities, traditionally upheld mainly by state institutions: (1) articulate a set of shared priorities for the society; (2) ensure the consistency of the objectives; (3) ensure the resources for implementing the objectives and; (4) to be able to account for what has been done:

– articulate a set of societal priorities. The first, and perhaps the most important, task for governing: articulate a set of priorities and objectives that are acceptable and accepted by the society. From the start, this puts public policies at the center of the system. Indeed, the exchange mechanisms employed in the private sector, and in civil society, assume that there is a national framework for the implementation of objectives which are sometimes complementary, but often competing;

– ensure coherence. This is the role of the government regulator, an important player for balancing concerns. This is done with greater or lesser consultation or association (according to the type of governance) with actors of the private sector and civil society;

– direct. Once the priorities and objectives have been set, it is necessary to know how they will be achieved, using what kind of control? The classic approach is to employ public instruments such as regulations and economic incentives. However, the trend is moving toward instruments that include consultation and dialogue with the actors of the society;

– take account. Each of the actors must take responsibility for its actions and must be able to account for them. The complexity of the policies, the fragmentation of the political parties and the public's very limited capacity to endorse or reward a politician before the elections have taken place, make *ex post* evaluation of contemporary democratic systems particularly important.

To understand governance, therefore, requires the understanding of the nature of state/society relationships in the interest of the collective good. This requires the understanding of the political nature of governance. The four components above are fields on which these political exchanges have taken or can take place.

The ecosystem approach for integrated management of the sea and coastal areas involves a large number of players who have varied information. As with the old adage, "knowledge is power", global access to information, even for the smallest economic agents (e.g. small-scale coastal fisheries), enables players to acquire the information necessary to better negotiate their rights. At the heart of the approach are monitoring and evaluation tools for the entire procedure and for the role played by each group of players. Their information and participation can greatly help in decreasing corruption. The media and politicians, who are in a position to ask for clarifications concerning decisions on the extraction of resources, can also do this. Another tool for integrated management is the strategic planning of maritime space. It aims to coordinate and optimize the use of maritime space to benefit development, and the coastal and marine environment. This decision-making tool creates a favorable climate for investment and for the creation of employment in the maritime domain. It does this at a national scale and across borders, between states that share a regional sea.

The Conference of the parties (CP10) of the Biodiversity Convention, held in October 2010 in Nagoya, Japan, passed some historic agreements. These related to the adoption of a new 10-year strategic plan (2011–2020), the mobilization of the resources necessary for the development of measures and the signature of a new international protocol on the access to and sharing of benefits from the genetic resources of the planet (the ocean constituting 90% of its essential volume). What may be seen, in the future, as even more historic is that, behind these agreements, the CP10 event brought together proponents of conservation and biodiversity with marine activities' administrators and players. We might, therefore, imagine that conservation tools, such as marine protected areas, may become tools for maintaining ecosystem services and thus underpin the sustainable development of maritime activities. The more we understand the complexity of coastal and marine socio-ecosystems, the more we can refine and combine the use of conservation tools for human well-being, in terms of innovation (economic competitiveness, growth and wealth), quality of life (health, education and safety), politics (citizenship and social debate) and culture (cultural heritage, communication/dialogues, etc.).

The 11th "Aichi" goal states that, by 2020, at least 17% of terrestrial and inland water, and 10% of coastal and marine areas, especially areas of particular importance for biodiversity and ecosystem services, are conserved through effectively and equitably managed, ecologically representative and well-connected systems of protected areas and other effective area-based conservation measures, and integrated into the wider landscapes and seascapes.

Box 7.3. *Aichi goal number 11: to join biodiversity conservation with sustainable development*

7.7.1. *National strategies for integrated management of the sea and coastal areas*

7.7.1.1. *Coastal zones and large marine protected areas*

Integrated management projects have appeared all over the world, but have been more coastal than maritime, and more local than national. One study [SOR 02] has helped to identify more than 700 projects in more than 90 countries. One of the recurring features of these projects is the establishment of MPAs. Around 6.3% of the world's territorial waters are now protected. This shows, despite everything, the considerable effort that has been made in this field. Nearly every country now has at least one MPA, and quite a few have established a national network of MPAs. Recently, the creation of large MPAs has considerably increased the ocean areas under protection. These include the Phoenix Islands of Kiribati, the Papahānaumokuākea national marine heritage site to the North-West of Hawaii, the Chagos Islands MPA of the United Kingdom, as well as the very recent natural marine park that covers the entire EEZ of New Caledonia. Ambitious initiatives, such as the Micronesia Challenge, the Caribbean Challenge and the Coral Triangle Initiative, are also signs of this positive trend toward the use of many MPAs, forming networks for biodiversity conservation and the protection of vulnerable ecosystems. MPAs are being considered increasingly less in isolation, but as tools that complement other forms of "managed marine area", whatever it may be. This is precisely the spirit behind the 11th Aichi objective, created at the 10th Conference of the Parties of the Biodiversity Convention at Nagoya (2010)[12].

7.7.1.2. *The continuity between the river basin and the maritime space*

One of the major problems for coastal waters is the deterioration of their quality due to sedimentation and pollution. The global programme of action (GPA) for the protection of the marine environment against land-based pollutions gives guidelines for actions to be carried out. These aim to achieve the objectives set in Johannesburg (2002), and those set in 2005, for the integrated management of water resources.

The GPA is targeted at governments, so that they might integrate the ecosystem approach into their public policy. They may also use this in regional partnerships, or international agreements, where river basins are shared. Despite repeated demands to link these policies with those in coastal zones, these initiatives remain disconnected with the upstream initiatives in the river basin. Some progress has nevertheless been observed in the control of eutrophication, beginning with a decrease in the quantities of nutrients, especially nitrates. At a regional scale, notable results have been made in the inland sea of Seto (Japan). Decreased nutrient quantities have also been measured in the Danube, along with a decrease in "dead zones" in the Black sea. The Global Environmental Facility (GEF) has encouraged, with some success, cooperation between the management of cross-border basins and

12 CBD CP10 – *10th Meeting of the Conference of the Parties to the Convention on Biological Diversity*, Nagoya, Japan, 18–29 October 2010. *Decision X/2 Strategic Plan for Biodiversity, 2011–2020.*

that of large marine ecosystems. This is the case, for example, for the Guinea Current strategic action plan.

7.7.1.3. *Economic exclusive zones*

It also appears that there are a number of countries that have incorporated the principles of integrated management of oceans into the management of their EEZ. This is true for the European Union and now a large number of its member states. They have all adopted an integrated maritime policy which covers all of the subjects and the fields of these maritime spaces' sustainable development.

The Ocean Policy Summit (2005) showed that all these integrated maritime policies coincided considerably on their general principles. They all recognize the necessity for transparency, the participation of players and the public, incentives for collective action and of forums for coordination with clearly defined responsibilities. Among those countries now involved are Brazil, Mexico, Jamaica, Canada, the United States, China, the European Union as a political body, as well as Portugal, Germany, France, Great Britain, the Netherlands, Norway, Russia, India, Japan, the Philippines, Vietnam, Australia, New Zealand and the list goes on.

In 2008, some 40 countries [CIC 08] had adopted an integrated maritime policy. Now, the ecosystem approach and integrated management of the sea and coastal areas are integral parts of these policies.

7.7.1.4. *Implementation of the ecosystem approach and integrated management in cross-border maritime basins.*

"The maritime basin is a relevant governance framework on which it is possible to define and develop a maritime policy. It is a space that brings together the different environmental, economic, social and political challenges; a space for cooperation between maritime and terrestrial policies; a space for international cooperation between coastal countries and the maritime basin. It is defined as a maritime space along with the coastline that it borders" (extract from the final report of the work group IV, debates on the Sea Grenelle in France).

The intergovernmental frameworks for regional maritime cooperation have existed for more than 30 years, with the regional conventions and the "Regional Seas" convention of the United Nations Environmental Programme

(UNEP), created in 1972. However, it is not sure what impact these cooperation provisions have had. The majority of these regional programmes have now incorporated the principles of the ecosystem approach and integrated management. This was first formalized by a protocol for integrated management of coastal areas, in the Mediterranean. However, it is evident, from the few regional assessments that have been carried out [SHI 09], that the impact of these integrated management projects remains modest. This is because of their isolation, their very small scale and the weak links that they create between environmental protection and local development; the latter obviously being the main priority for local populations.

7.7.1.5. *The European case*

In 2007, following consultations broadened to include all member states, the European Commission published a Blue book on the integrated maritime policy of the European Union and its action plan. The environmental pillar of this policy is represented by the recent "Marine Strategy" framework directive, adopted in 2008[13]. This directive is clearly based on an ecosystem approach for the management of human activities. It aims to obtain and maintain the good environmental condition of all European waters by 2020. The directive stipulates that it should be transcribed into the national law of all member countries by 15 July 2010. It recognizes that decisions can no longer be made only considering each sector, but should be made at the level of cross-border marine ecosystems (regional seas); the preservation of these being the main condition for sustainable development of maritime activities. This, therefore, requires the development of an approach that interlocks activities at various levels, from local to regional (regional seas). This should connect the member states, so that those that share a maritime basin collaborate with each other. This assumes that each country has its own integrated maritime policy. This is underway for most of them.

Among the three key instruments[14] for the implementation of the integrated maritime policy of the European Union is Maritime Spatial Planning. This approach, which has been the subject of numerous scientific

---

13 Directive 2000/56/CE, of 17 June 2008, establishing a framework of community action in the field of water policy, OJ L 164/16.

14 We should remember that, in the action plans of maritime integrated policy, maritime strategic planning is one of the three preferred instruments for implementation; the two others being the networking of monitoring systems at sea, and the European Marine Observation and Data Network (EMODNET).

publications and guides, is supported by a road map adopted by the European Commission in November 2008. It contains 10 principles briefly presented in Box 7.4.

1. Using MSP according to area and type of activity

The first principle gives some flexibility in the implementation of MSP. It does not necessarily have to cover an entire maritime area, or at least does not have to apply in the same way according to the area: more regulatory in highly used areas (as is already the case for navigation passages) or vulnerable areas (Natura 2000, marine parks); it should be more of a guide in less used areas.

2. Define objectives to guide MSP

This principle, which is only quickly described in the road map, deserves further clarification: planning helps to form a management framework, based on a vision and strategic orientations derived from assessments of good environmental status, as stipulated in the "Marine Strategy" framework directive. MSP is, therefore, a tool for implementing the strategic choices negotiated between players and run by the public authorities.

3. Develop transparent MSP

This is closely connected to the last point: it is important that all the players participate in the entire governance process, for which MSP is one tool.

4. Participation of the players

This principle somewhat sets straight the vision of an MSP which would not concern itself with coordinating with the coastline, since it touches upon coastal regions and their players. This principle raises the issue of coordinating the different available instruments.

5. Coordination among Member States

New forms of governance should be invented during the decision-making process, for the widest possible acceptance. This would lead to a more coordinated, and thus more effective, implementation of the action plans or the projects resulting from this process. In this context, the coordination of public policies is essential, especially through interministerial coordination

6. Guarantee the legal scope of MSP at a national level

The conditions for the application of this principle refer back to principle 1. This means that this principle depends on the intensity of the maritime uses in a given region or a maritime basin. Aside from MSP, the appropriate management framework should be ascertained at a more global level: of the governance and integrated management of the sea and coastline.

7. Cross-border cooperation and consultation

MSP must be used in accordance with the "Marine Strategy" framework directive. Agreements should be made between states to coordinate the development of their maritime activities within the same maritime basin or marine ecoregion. Regional conventions are the preferred form for developing these agreements, since they provide tried and tested mechanisms for coordinating between states. In this respect, the situation in the Mediterranean is more complex since it involves several non-member countries. The Mediterranean Union could help to push forward the Barcelona Convention in this field as well.

8. Monitoring and evaluation of the planning process

The requirement for a monitoring and evaluation system is not unique to MSP; it should already be part of the action plans for integrated management of the sea and coastline.

9. Coherence between land-use planning and MSP

This principle refers to the interconnections between the river basin, coastal area and open seas, and the necessity to coordinate the management of these environments. This requires the coordination of legal frameworks and ensuring that planning instruments are compatible.

10. Constant input of data and knowledge

This principle raises the issue of coordinating national databases (which themselves need to be structured in a network) with the European databases, currently under construction.

Box 7.4. *The 10 principles of the road map on strategic planning of maritime space (MSP)*

Before proceeding to strategic planning of maritime space, a major ambiguity must be removed: land and sea spaces are considered separately, although at least 80% of pollutant input comes from the river basin. Furthermore, maritime activities require bases on land (e.g. port sites) in order to develop. The key for effective maritime strategic planning is thus found at the land–sea interface. Such a coordination cannot occur only within the context of maritime policies created for integrated management of the sea and coastlines – i.e. frameworks for ensuring the coherence of the processes (ecosystem approach, integrated management of coastal areas, etc.) and tools such as strategic planning of maritime space. In Europe, this is achieved by the interaction of the Water Framework Directive (WFD) – which concerns the river basin, up to the coastline – and the Marine Strategy Framework Directive (MSFD), which extends the goal of good environmental status out to the sea.

After several debates, stirred up by the rivalry between the European Commission's Directorates General Environment and "MARE", a directive was adopted, in July 2014, aiming to create a common framework for the planning of maritime space in Europe. Although coordination with integrated management of coastal areas is mentioned in it, it would undoubtedly been much more simple to standardize the approach under the name of integrated management of the sea (EEZ) and coastal areas. This would have sent a clear message to all of the decision-makers, administrators and stakeholders.

7.7.1.6. *Large marine ecosystems and the Global Environment Facility*

The GEF has played a central role in financing, preparing and implementing at least 16 large marine ecosystems projects. These have involved more than 100 countries, and have covered issues such as the overexploitation of fishing stocks, the decline in the trophic level of catches, the loss of habitats and coastal pollution. Each of these projects has used the principles of the ecosystem approach, and integrated management, while developing relevant indicators for these. [WOW 07] has made a first assessment of this integration.

Alongside the seascapes of the NGO Conservation International's (CI), the marine ecoregions of the World Wildlife Fund (WWF) and The Nature Conservancy, GEF/GEM projects constitute the largest implementation of the ecosystem approach for integrated management of the sea and oceans. They have helped to identify and exchange a set of practices. However, they have not yet managed to establish an ongoing dialogue between the 2,500 "practitioners" currently involved (to a greater or lesser extent) in these projects. The aim is to increase this number to 10,000 practitioners in 2012 so long as the GEF continues to fund the initiatives. Based on this wealth of experience, a guide for the governance of the socio-economic development of large marine ecosystems (LMEs) was published in 2006 [OLS 06].

7.7.1.7. *The Regional Seas Programme*

The UNEP's Regional Seas Programme involves 18 maritime regions and mostly concerns subjects related to the maintenance of biodiversity and ecosystems' health, land and marine pollution, MPAs, and the integrated management of coastal areas. Twelve "regional seas" programs have adopted legally binding agreements, along with specific protocols for the aforementioned fields, and action plans, sometimes grouped together under the umbrella of a regional strategy for sustainable development. Most of them have adopted the principles of integrated management of the sea and coastline, sometimes with a set of indicators to measure progress.

Although it is not a "regional seas" programme, it should be noted that the HELCOM (Baltic Sea) and OSPAR (North-East Atlantic/North Sea) conventions have also used the principles of the ecosystem approach for integrated management.

In 2008, the signatories of the Barcelona Convention signed the "Protocol on Integrated Coastal Zone Management in the Mediterranean". This will take effect as soon as six countries ratify it (there were five in October 2010). Even before it has been put into action, it is already being held as a model for other regions. It is being considered for use as part of the Nairobi Convention, in the Southwest of the Indian Ocean.

7.7.1.8. *Other initiatives*

On a more global scale, the 9th Conference of the Parties (May 2008) of the Convention on Biodiversity (CBD) adopted criteria for identifying ecologically or biologically important areas, in the open sea and deep waters, which require protection. It also established guidelines for establishing representative networks of MPAs.

In 2009, the CBD organized an experts' workshop[15] on these issues in order to help governments in its application. The recommendations resulting from this were submitted to the Conference of Parties (COP10), held in Nagoya, Japan, in October 2010. This was the starting point for the organization of regional workshops covering all oceans and regional seas. Due to these, for the first time, regional fishing organizations collaborated

---

15 *Expert Workshop on Scientific and Technical Guidance on the Use of Biogeographic Classification Systems and Identification of Marine Areas beyond National Jurisdiction in Need of Protection*, Ottawa, Canada, 20 September–2 October 2009.

with regional conventions for environmental protection, in an attempt to delineate ecologically or biologically important areas in international waters.

Several countries or groups of countries, such as the United States and the European Union, have integrated the principles of the ecosystem approach into their fisheries policy. This is particularly true for cross-border fish stocks, and for highly migratory species. This is recommended in articles 5 and 6 of the 1995 UN Fish Stocks Agreement (UNFSA), and the FAO's Code of Conduct for Responsible Fisheries.

Regional fisheries management organizations – such as the Western and Central Pacific Commissions, the South East Atlantic Fisheries Organization and the new South Pacific Regional Fisheries Management Organization – have all integrated the concepts of ecosystems and biodiversity conservation. However, the oldest organizations, such as the North East Atlantic Fisheries Commission, have not yet revised their founding text in this regard.

7.7.2. Implementation of the ecosystem approach for integrated management of areas beyond national jurisdictions

Governance of the 64% of the oceans that make up international waters is a major issue that countries must consider and discuss over the next decade. This is an area in which many maritime activities take place. These are sources of profit for regional and global economies. Despite the progress made at national and regional (regional seas) level, governance of international waters remains largely sectorial in nature. It is, therefore, difficult to deal with the interconnected issues, such as fishing, the extraction of genetic resources, maritime transport, pollution, offshore oil and gas, scientific research, climate change, or carbon sequestration and storage. Moreover, there is much conflict in the views of developed and developing countries, and between industrial and environmental interests, as to how to best improve the governance of this shared part of the world.

In the case of deep-water habitats, such as hydrothermal vents, it is not known what impact certain uses might have on the structure, function and properties of these ecosystems. The ecosystem approach is not only concerned with the direct and indirect value of uses, but also with the intrinsic value of ecosystems that might be immediately threatened by exploration and bio-prospection activities. The same is true for mining or the use of hydrogen in this energy sector.

It is still too early to implement strategic maritime planning in these international regions. However, large oceanic MPAs have attracted increasing attention and can be regarded as the first components of an ecosystem approach. For example, negotiations in the North East Atlantic have led to the protection of a part of the Charlie Gibbs fracture zone (deformation zone in the mid-Atlantic ridge). This involves the OSPAR regional convention and the Regional Commission for Fisheries covering the North East Atlantic. Similarly, the International Seabed Authority (ISA) works with its "contractors" and the international community to develop environmental action plans for granting mining exploration permits in international zones.

According to articles 204–206 of the UN Convention on the Law of the Sea, states are required to assess the potential effects of activities within their jurisdiction or control that may cause pollution or significant damage to the marine environment. The results of this work must be communicated to international organizations and made available to all states. The same applies for the Convention on Biodiversity (articles 4 and 14), which requires that signatories carry out environmental impact studies when their activities might affect biodiversity in areas under national jurisdiction, as well as in international waters. Nevertheless, progress remains to be made to better specify how these studies should be carried out in these waters. Other activities that are subject to international resolutions include deep-sea trawling, and the discharge of waste into the sea.

These examples show that a complex set of international and regional legal instruments already exists for the conservation and management of international waters. However, as already stated before, there is no coherent governance system which might ensure the assessment and regulation of new activities that might endanger marine ecosystems [CJE 08]. For this reason, organizations, such as the International Union for Conservation of Nature (IUCN), demand that the UN General Assembly adopts, as a matter of urgency, a resolution urging states to: (1) develop assessment procedures, including for the cumulative impacts of human activities on marine biodiversity; and (2) to ensure that the activities under assessment, which may affect the environment, are subject to prior authorization for the nationals and vessels for which the states are responsible, while ensuring that they are managed in such a way as not to cause damage to the environment.

In spite of all the efforts, the international waters management approach remains fragmented. It is currently not able to meet the challenges and threats it faces, especially due to the effects of climate change. Would a new United Nations Environment Agency be able to create this coherence? It would be probably not be enough, and we should probably head toward a new organization for the global governance of oceans.

All the marine zones defined by the Convention of the Law of the Sea, except for the (12,000 coastal) territorial waters, have the status of "common heritage of mankind". This includes all the biological and mineral resources that lie outside of territorial waters. Coastal states keep their rights to use the resources within the EEZ, as well as the mineral resources of the continental shelf. However, they are responsible for maintaining a good environmental quality, by using these resources sustainably. The usage rights thus come with obligations to report to the new United Nations' Global Ocean Organization (UN-Global Ocean). The International Seabed Authority and the Continental Plate Commission have been integrated into UN Global Ocean. The regional fisheries management organizations have been integrated into the regional organizations of marine management, which looks after the sustainable management of all the resources on the high seas. It also deals with the management of stocks of cross-border and highly migratory species in cooperation with the relevant coastal states. UN Global Ocean takes care of the global governance of oceans and ensures that the rules for their conservation and sustainable use are adhered to. UN Global Ocean has legal means of action for this, via the International Tribunal for the Law of the Sea (located in Jamaica). Regional cooperation between coastal states continues in the form of agreements and joint programmes.

Box 7.5. *What shape could tomorrow's ocean governance take? (from [WBG 13])*

7.7.3. *Hurdles to overcome*

While much has been achieved since the Rio Conference (1992), this has been very fragmented. The issues of coordinating between institutions, and between policies/programmes, have largely undermined the effectiveness of this approach, which is, by definition, coordinated with the integrated management of the sea and coastal areas.

Scale of issues	Time scale	Relevance of the approach for resolution/prevention of the issues	Relevance of the approach for adaptation to regional/global change	Intervention frameworks	Comments
Local (land/sea interface)	Present	Strong if coordinated between intervention sites	Weak (no monitoring of local players)	Land-use planning/strategic planning	The geographic area of the policies can vary widely
Regional (river basin and/or maritime basin)	Present	Strong if it helps to link management of the river basin and the maritime basin	Depends on local players and existing arrangements for governance	Regional conventions – integrated management of river basins/integrated management of large marine ecosystems	The concept of cross-border maritime basin will encourage more participation from local players
Global	Present	Strong if inter-regional coordination, and the disparities between North/South development are taken into account	Strong if included in local management plans	Global governance and international conventions/agreements	The effects of global change separate even more the rich countries of the North and the poor countries of the South
Historical heritage (varied geographic scales)	Past	Strong if taken into account	Very strong: adaptation is the only effective strategy	None or several	Greenhouse gas emissions remain as they are due to the lack of alternatives

Table 7.2. *Application of the ecosystem approach for integrated management of the sea and coastlines, with spatial and temporal scales for the resolution of problems (adapted from [MEE 09])*

How does the current situation differ from that of the past?

– Up to now, practices have led to the accumulation of impacts on the climate, habitats and biodiversity, seriously disturbing the viability of services rendered by the ecosystems.

– The social and economic pressures on coastal systems are increasingly on a regional and global scale, over which local populations have little power.

– The vulnerability of coastal systems is increasing, in the face of natural catastrophes as well unregulated global markets.

– With an ever-increasing world population, migrations from disaster areas are no longer sustainable; however, the effects of climate change will be felt most severely by the southern nations, which are the poorest.

Therefore, it is essential to take account of globalization in order to resolve the issues which marine and coastal systems are facing. In order to resolve complex issues, we must understand the processes which drive them and ensure that systems of governance are not only adapted, but also coordinated between the different levels of intervention.

Current local practices of integrated management should be incorporated into wider approaches, able to take into consideration changes at higher levels, which might directly or indirectly influence marine and coastal systems. Among these, global change and climate change are the most difficult to comprehend. However, the bigger they become, the greater the need for global governance.

7.7.4. Size and limits of global expertise

The creation and evolution of all of these policies and programmes depend on a network of international organizations – the United Nations, development banks (including the World Bank), the large international NGOs – as well as multiple public and private agencies, formal or informal institutions and "global" experts. These experts build the regulatory instruments and devise the norms of this "empire" of "environmental knowledge" [GOL 04]. Experts and expert knowledge (published by the experts) are unquestionably the core features of the emergence of globalization. This is through the dissemination of global means of conceptualization, the transformation of the populations' experiences and the behaviors, the spreading of "environmental know-how" and the sustainable development paradigm. These can be communicated through governmental, private or group networks, from a global to a local level (but not in the opposite direction). Scientific knowledge and expertise are important factors for the emergence of globalization, via the creation and dissemination of conceptual models and a global system.

These models have created an ontological link in between overarching environmental and social issues. The power of these scientific authorities is expressed through the development of databases and monitoring networks, which provide sophisticated systemic models of the evolution of global processes, and remodel local places through given categorizations. In this regard, local knowledge is not ignored, but exploited for the purpose of globalization and abstract modeling. The standardization of models and solutions hampers the expression of situations and practices in the field which, like biodiversity conservation ensure the resilience of our planet.

7.8. Conclusion

From recreation space to heritage place, the coastline has become a common asset of mankind. Initially confined to a small perimeter – coastal areas, shorelines and the waterfront – the seaboard stretches out into the maritime domain, the last frontier not only between states, but between universes. It is the site of all interconnections and its governance has become a global challenge. The issues of scale take on great importance for geopolitical, strategic and environmental reasons. This is particularly true for global threats that weaken the ecosystems, and the new resources essential to the sustainability of the only system current working on the planet: the liberal capitalist system, focused around law and the economy (the market). On various levels (regional, national, European and international), an epistemic community is seeking, through UN institutions, to find cross-border systems of sea and coastal governance. In this way, they might establish the necessity of a "New Deal" at a global scale for the maritime, economic, social and environmental domains.

The Montego Bay Convention on the Law of the Sea (adopted in 1982, enforced in 1994) was the first international ruling and true constitution on the sea. It represented a compromise between the principles of freedom, sovereignty and common heritage of mankind. Nevertheless, this compromise is in discrepancy with contemporary issues, and does not link with other international schemes established in the wake of growing environmental concern and the new scientific knowledge in biology, climate (such as climate change), pollution, biodiversity conservation and risk management. In the future, ocean global governance must be able to make use of a set of integrated legal tools and institutional arrangements such as the creation of a Global Ocean Organization as mentioned above.

Nature and society, from a local or national perspective, are nowadays often addressed using global themes (biodiversity, climate, bioinvasions, pollution, etc.). Globalism represents the explicit framework for structuring nature and society, while declaring the neutrality and objectivity of the epistemic community (scientists and experts). In order to avoid standardizing problems and solutions, it is necessary to rethink how knowledge is gained. Local knowledge must be taken seriously, be considered as important as "learned knowledge" and not simply exploited in the name of "good governance". It is not a question of contrasting local with global, but of shedding the simplistic vision of knowledge and expertise; and thus gaining a "situational", contextual knowledge, using existing meeting places or to create them if they do not already exist. This attitude requires a more open view of science, a demystified philosophy [STE 13] and an acceptance of knowledge-bases and methods different to one's own. It leads to collaboration between disciplines, with players and their local groupings. It sees the population and its representatives as sources of information and whistleblowers, involving them in building knowledge.

"Good governance" in the implementation of multi-level integrated management of the sea and coastal areas is a case of working to reduce the gap between ecological considerations and democratic process. Governance is not an issue of optimums, effectiveness and transparency. It is first an issue of democracy, for creating democracy. It is not so much about defending tangible outcomes (although they are essential to testing) but, first and foremost, defending the processes. We must make sure that the schemes put in place create democracy. The main issue is "how to ensure that nature and animals are a part of the political community", while accounting for the humans' various positions and abilities to choose.

We are, therefore, far from a fixed "technical" definition of integrated management. We are much closer to an integrated management underpinned by a "cosmopolitical" vision, supported by systems of governance where adaptive management and social learning might mix. The creation of a shared world requires us to reconsider our modern perception of the world, which is based on dualisms between object and subject, nature and society, equality and justice, modern and non-modern, etc. In this respect, knowledge should not be disconnected from man's plans and hopes to learn how to live together, to achieve personal fulfillment and to fulfill the moral development of communities. It should lead to a better understanding and thus a good ecological and environmental status.

7.9. Appendix: some proposals for global governance of seas and coastal areas

7.9.1. *Strategic requirements at national and local levels*

– Extend national policies and programmes to territorial waters and to all of the EEZ as an integrated maritime policy.

– Devise and implement, with the necessary financial means, laws and regulations specific to the coastline and maritime area under national jurisdiction. The ministers involved should coordinate between themselves, in particular by creating a parliamentary commission dedicated to the coastline and sea.

– Initiatives for the reduction of and adaptation to the effects of climate change must be assimilated into the plans for integrated management of the sea and coastlines, and not treated separately. Training of the officials involved is essential.

– Share experiences and good practices concerning the ecosystem approach and integrated management. In this regard, the creation of a network is essential: from a scientific and operational aspect in the field, and also for the communities and departments responsible.

– Devise incentive frameworks such as certifications of good practices for the ecosystem approach and integrated management. The Partnerships in Environmental Management for the Seas of East Asia (PEMSEA) has already launched one such initiative with local communities[16].

– Develop and strengthen over the long-term abilities for governance and management of coastlines and territorial waters. The recommendations of work group 3 of the Sea Grenelle relate especially to the improvement of skills of maritime administration and the development of social and environmental responsibility with the economic players. Generally speaking, the goal is to strengthen the abilities of individuals and organizations by skill transfer, to familiarize them with the key concepts and to help them develop a systemic vision.

– All of these actions require increased financial commitments. The United Nations Commission for Climate Change estimates that the cost to adapt to the effects of climate change is about 11 billion dollars/year, not

---

16 www.pemsea.org/programmes-and-projects/scaling-up-icm.

including the effects due to the increased intensity of climatic events, such as storms, tornados, rainfall, etc.

7.9.2. *Strategic orientations at a regional level*

– Promote and strengthen the key roles played by regional seas programmes and the organizations linked to the development of LMEs approach; the goal being to synchronize national initiatives with cross-border ones. In particular:

- increase the current number of 2,500 practitioners of LMEs management to 10,000 in 2012, in order to strengthen the "community of practice" of the GEF in its knowledge fields and its operational application in terms of management;

- ensure GEF continuing support to LME projects so that they become financially independent one day;

- use the UNEP/LME reports as bases for regular evaluation (every 3 years) of the state of marine ecosystems in the world.

– Promote and apply, at a national level, the regional protocols for integrated management of coastal zones, as in the Mediterranean;

– Encourage and implement multiple, but coordinated, ecosystem approach and sea and coastline integrated management projects. This can be done through various regional organizations, such as the bilateral fisheries commission, the regional fisheries organizations and the other relevant sectors.

7.9.3. *Strategic operations for areas outside of national jurisdiction*

– Apply the same principles of the ecosystem approach and integrated management.

– Develop capabilities in the fields of environmental evaluation, planning and in the means of governance for decision-making.

– Develop institutional capabilities to manage the interactions between uses and their effects on biodiversity, and the ecosystem services.

– Develop capabilities for tracking and monitoring and the implemented practices.

– Necessary financing to support the management actions.

If the international community agrees fairly easily on the concepts, there is still major disagreement on the way to implement them and which institutions would be responsible for this. In any case, agreements on the following points are urgently needed:

– the international mechanisms to designate MPAs in the areas outside of national jurisdiction, either based on existing institutions or using new ones;

– the norms and procedures for environmental impact studies on new activities or the growth of existing activities;

– the principles that apply to the management of oceanic resources, in particular the principles of the ecosystem approach;

– the choice of an international organization for the coordination of the various agencies involved in the management of oceanic resources.

7.10. Bibliography

[BRU 87] BRUNDTLAND G.H., *Our Common Future*, Oxford University Press, 1987.

[CHA 99] CHATEAURAYNAUD F., *Les sombres précurseurs: une sociologie pragmatique de l'alerte et des risques*, EHESS, 1999.

[CIC 08] CICIN-SAIN B., VANDERZWAAG M., BALGOS M., Nippon Foundation Research Task Force on National Ocean Policies, 2008.

[CLA 03] CLARK W.C., DICKSON N.M., "Sustainability science: the emerging research program", *Proceedings of the National Academy of Science*, vol. 100, no. 14, pp. 8059–8061, 2003.

[CUR 08] CURY P., MISEREY Y., *Une mer sans poissons*, Calmann-Levy, 2008.

[GAT 14] GATTUSO J.P., HANSON L., GAZEAU F., "Ocean acidification and its consequences", *Ocean in the Earth System*, ISTE, pp. 131–185, 2014.

[GJE 08] GJERDE J., Regulatory and Governance Gaps in the International Regime for the Conservation and Sustainable Use of Marine Biodiversity in Areas Beyond National Jurisdiction, International Union for Conservation of Nature Marine Series 1, 2008.

[GOL 04] GOLDMAN M., "Imperial science, imperial nature: environmental knowledge for the World (Bank)", in JASANOFF S., MARTELLO M.L. (eds), *Earthly Politics: Local and Global in Environmental Governance*, MIT, Cambridge, 2004.

[HOL 86] HOLLING C.S., "The resilience of terrestrial ecosystems: local surprise and global change", in CLARK W.C., MUNN R.E. (eds), *Sustainable Development in the Biosphere*, Cambridge University Press, 1986.

[KAL 99] KALAORA B., "Global expert: la religion des mots", *Ethnologie Française*, vol. XXIX, no. 4, pp. 513–527, 1999.

[KAL 10] KALAORA B., *Rivages en devenir: des horizons pour le Conservatoire du Littoral*, La Documentation française, 2010.

[LEV 07] LEVREL H., *Quels indicateurs pour la gestion de la biodiversité*, Les Cahiers de l'IFB, 2007.

[MEA 05] MEA, *Ecosystem and Human Well-being: Synthesis*, Millennium Ecosystem Assessment, Island Press, 2005.

[MEE 09] MEE L.D., "Between the devil and the deep blue sea: the coastal zone in an era of globalization", Background Paper 1, LOICZ Dahlem-type Workshop, *Global Environmental Change in the Coastal Zone – A Socio-Ecological Integration*, Skjetten, Norway, 15–19 June 2009.

[MON 14a] MONACO A., PROUZET P., *Vulnerability of Coastal Ecosystems and Adaptation,* ISTE, London and John Wiley & Sons, 2014.

[MON 14b] MONACO A., PROUZET P., *Ocean in the Earth System*, ISTE, London and John Wiley & Sons, New York, 2014.

[OLS 06] OLSEN S.B., SUTINEN J.G., JUDA L. *et al.*, *A Handbook on Governance and Socio-Economics of Large Marine Ecosystems*, CRC Press, University of Rhode Island, 2006.

[OST 10] OSTROM E., *Gouvernance des biens communs. Pour une nouvelle approche des ressources naturelles*, Etopia/De Boeck, 2010.

[SEN 12] SENAT, Rapport d'information du groupe de travail sur la maritimisation, no. 674, 2012.

[SHE 10] SHERMAN K., Adaptive Management Institutions at the Regional Level: The Case of Large Marine Ecosystems, NOAA, 2010.

[SHI 09] SHIPMAN B., HENOCQUE Y., EHLER C., The way forward for the Mediterranean Coast. A framework for implementing regional ICZM policy at the national and local level, SMAP III/2009/ICZM-PR/ENG Priority Actions Programme Regional Activity Centre, Split, June 2009.

[SOR 02] SORENSEN J., Baseline 2000 background report: the status of integrated coastal management as an international practice (second iteration), 26 August 2002.

[SPA 07] SPALDING M.D., FOX H.E., ALLEN G.R. et al., "Marine ecoregions of the world: a bioregionalization of coastal and shelf areas", *Bioscience*, vol. 57, no. 7, pp. 573–583, 2007.

[STE 13] STENGHERS I., *Une autre science est possible! Les empêcheurs de penser en rond*, La Découverte, Paris, p. 211, 2013.

[WBG 13] WBGU, World in transition: governing the marine heritage, German Advisory Council on Global Change, 2013.

[WOW 10] WOWK K., Achieving ecosystem management and integrated coastal and ocean management in regional ocean areas, Working Paper Series on Meeting the Global Goals of Achieving Ecosystem management and Integrated Coastal and Ocean Management by 2010 in the Context of Climate Change, 2010.

Ocean Industry Leadership and Collaboration in Sustainable Development of the Seas

8.1. Ocean industry sustainability: challenges and opportunities

Leadership and collaboration by the diverse, international ocean business community are essential in addressing ocean sustainability issues and maintaining industry access and social license for the responsible use of marine space and resources.

Sustainable use of the dynamic, interconnected global ocean presents unique opportunities and challenges for ocean industries. Overall, as the health of the marine environment declines, ocean industries are often held responsible for their impacts on the ocean by the public, governments, non-government organizations (NGOs) and inter-governmental organizations (IGOs). Ocean stakeholders are pushing for increased governance in a variety of international venues where international ocean rules are established. Some of the most important ocean governance developments are being pursued through non-sector-specific international policy processes that include oceans, e.g. the Convention on Biological Diversity (CBD) and the UN Convention on the Law of the Sea (UNCLOS), etc. Balanced and comprehensive information on industry efforts to address marine environmental issues is often not seen in these processes, and there is a need

Chapter written by Paul HOLTHUS.

for strategic and coordinated industry participation. Marine industries are often only portrayed as the cause of ocean problems, and it is difficult to correct this misperception if they are not engaged in ocean sustainable development efforts with other stakeholders.

As a result, private sector access to ocean resources, services and space – even by companies with the best environmental record – is increasingly at risk due to the loss of social license to operate in the seas. Responsible ocean companies are making efforts to do business more sustainably. However, the efforts of one company, or even of a whole sector, are not enough to address the collective global impacts manifested by a diverse range of industries in a shared global ecosystem.

The private sector is now well prepared to develop and deliver solutions in response to society's demands that the marine ecosystem be exploited responsibly and that the industry impacts be minimized. A cross-sectoral ocean business community of leadership and collaboration is needed to address marine environmental issues, differentiate good performers, create collaboration with like-minded companies within and across sectors, and engage ocean stakeholders and policy processes. Given the size and scope of ocean industries, forward-looking companies and executives have a particular role to play in providing leadership in collaborative, industry-driven ocean sustainability.

8.2. Status and trends in economic use of marine space and resources

To understand the role of industry leadership and collaboration in ocean governance and sustainable development, it is critical to have a clear understanding of the status and trends in the economic use of marine spaces and resources, as well as the potential new types and areas of use. Achieving a balance between "blue growth" jobs, and a healthy marine environment requires that ocean industries address both the economic opportunities and environmental effects of their ocean activities. Improving ocean governance and sustainable development will require coordinated leadership and collaboration by the diverse ocean business community.

8.2.1. *Shipping*

International shipping traffic growth has been twice that of economic activity for the past 60 years, during which time world trade more than trebled, comprising 45% of global gross domestic product (GDP). In 2013 there were approximately 51,902 merchant ships of over 500 UMS[1] operating internationally. Globally, shipping is generally either as liquid cargo, e.g. oil, petroleum products, chemical or as dry cargo/bulk goods, of which the most important are iron ore, coal, grain, phosphates, bauxite, non-ferrous metal ores, feed and fertilizers. The most significant cargo worldwide is crude oil, which makes up about 25% of all goods transported by sea[2].

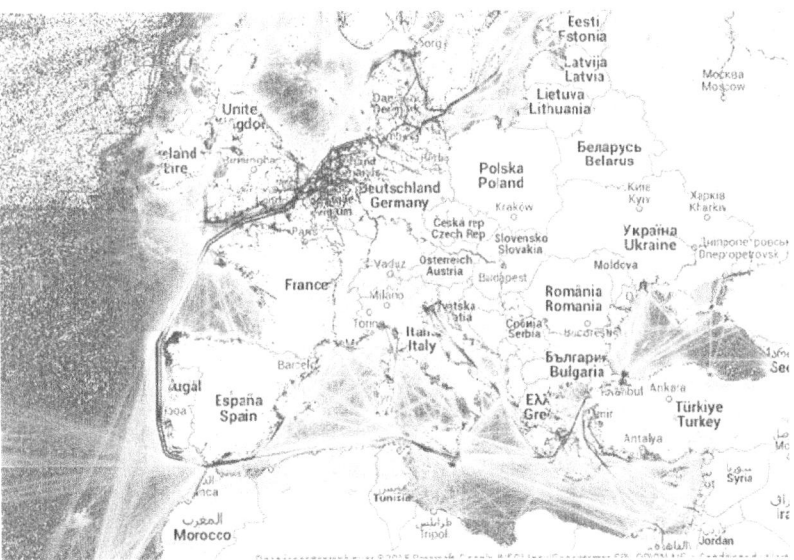

Figure 8.1. *Example of density and intensity of maritime traffic on European coasts (source: Marine Traffic 2015)*

Otherwise, most goods travel by container ship and, since 1985, global container shipping increased by about 10% annually, with about 137 million containers transported in 2008. According to the 2011 UNCTAD study on maritime transport, 1.477 billion tons have been transported by way of

1 UMS : Universal Measurement System , 500 UMS = 1415 cubic meters.
2 From UNCTAD Review of Marine Transport 2014.

containers, and in 2010 the number of FEU was 140 million or 1.3 billion tons [CNU 12][3].

From 1990 to 2010, transport via containers has had the fastest increase rate in maritime transport with an average growth of 8.2% over that particular period [CNU11]. There are a relatively small number of principal transport routes, and the busiest are the approaches to the ports of Europe, US and East Asia, particularly Japan, as well as Shanghai, Singapore and Hong Kong. Narrow straits concentrate maritime traffic, e.g. Straits of Dover, Gibraltar, Malacca, Lombok and Hormuz, and the Cape of Good Hope. The heavy traffic to Northern Europe and the Eastern US, and between these two areas, makes the North Atlantic an area of especially high shipping traffic, with its associated challenges (Figure 8.2).

Figure 8.2. *Container ship*

8.2.2. *Offshore oil and gas*

Offshore oil and gas industry fields explored in the past were relatively shallow and limited in size. Now, 45% of the 2.7 billion barrels of

---

3 FEU: Foot Equivalent Unit – defined a 20 feet or 40 feet container . The dimension, of a 20 feet container are (20*8*8 fees).

recoverable oil left is offshore, with energy firms gradually moving to deeper waters as shallow water reservoirs become depleted. By 2035, deep-sea production will almost double to 8.7 million barrels a day, driven by developments in the US Gulf of Mexico, Brazil, West Africa and Australia (mainly for gas)[4]. The Gulf of Mexico remains the world's most valuable deepwater province, despite the many recent large finds elsewhere.

Since the discovery of ultra-deep oil reserves under a thick layer of salt off the coast of Brazil, the offshore oil and gas industry is exploring ever deeper and drilling further under the sea bed, exploring the subsalt layers 7 km below sea level (below 2.5 km of ocean water, 3 km of rock and 2–3 km compacted salt). "Ultra-deep" wells, drilled in water and at least 1.5 km deep, now account for over half of all the world's new fuel discoveries. Addressing the technological and safety challenges requires significant capital, with investment in the global deepwater and ultra deepwater exploration and production market, expected to be worth 3.2 billion USD, in 2013, in an industry where a single offshore well may cost 70 million USD to drill. In a global fleet of over 1,200 rigs and drilling vessels, more than 80 rigs now have the ability to work in ocean depths of more than 2.5 km (Figure 8.3). This contrasts with the fewer than 10 in 2000 and double the number at work just 2 years ago.

Figure 8.3. *Oil platform*

---

4 Information from International Energy Agency: "Platform-free offshore oil could flow within a decade".

8.2.3. *Fisheries*

The world's most productive fishing grounds are largely confined to areas that make up less than 10% of the global ocean, often associated with areas of strong primary production of biomass in the oceans, i.e. continental shelves and upwelling areas (see Chapter 4 of [GRO 14]). Marine fishery catches increased from 16.7 million metric tons in 1950 (86% of total world production) to a peak of 87.7 million metric tons (MT) in 1996. Since then, global landings of fish and seafood have declined, with fluctuations reflecting the variation in catches from a few highly productive areas, particularly the Northwest and Southeast Pacific, which account for a large portion of pelagic species catches. In 2012, fishing-based production was estimated at 79.7 million in marine waters and 11.6 million in continental waters [FAO 14]. This equates to approximately 51% of the world's production of fish for human consumption between 2011 and 2015[5]. Based on average catches in the 2005–2009 period, the most productive fishery areas are the Northwest Pacific (25%), Southeast Pacific (16%), Western Central Pacific (14%), Northeast Atlantic (11%) and Eastern Indian Ocean (7%). All other marine fishing areas contribute less than 5% of the global total catch. The proportion of overfished stocks has increased from 10% in 1974 to 29% in 2011 [FAO 14].

Figure 8.4. *Fishing trawler (copyright Mike Markovina)*

---

5 Total estimated at 136 milliom tons (see Figure 4.1 in [GRO 14]), plus 22 million tons produced par industrial fishing for animal consumption.

The patterns of marine fishery landings differ over time. Some areas have oscillations in total catch but a declining trend is not evident. In the Atlantic, this includes the East Central and Southwest areas. Many others have a decreasing trend in catch; this includes four of the Atlantic fishery areas: Northwest (a decrease of 55%), West Central (down to/by 46%) and Northeast (down to/by 35%), with the Southeast decreasing somewhat less. Third, there are areas that have shown a continual increase in catch since 1950, although none of these in the Atlantic. In the high seas, migratory tunas and related species are the most valuable fishery resource, with production highest in the Pacific, followed by the Atlantic and Indian Oceans. The harvest of high-seas fishery resources increased from less than 0.5 million MT in the early 1950s to 5.5 million MT in 2006.

8.2.4. *Aquaculture*

Aquaculture provides half of the 15% of animal protein consumed globally [FAO 12]. Aquaculture has grown at 6.6% per annum, making it the fastest growing animal-food-producing sector – much faster than the 1.8% annual global population increase. While aquaculture production (excluding aquatic plants) was less than 1 million MT per year in the early 1950s, production in 2008 was 52.5 million MT, with a value of 98.4 billion USD. Aquatic plant production through aquaculture in 2008 was 15.8 million MT, with a value of 7.4 billion USD. Global aquacultural production between 2011 – 2013 was estimated at 66 million tons on average per year [FAO 14, OEC 14]. By 2030, aquaculture will account for 65% of fish protein production, but with a decelerated growth estimated at 2.5% during the 2013 – 2023 decade [GRO 14].

Figure 8.5. *Inshore and offshore pen-nets*

World aquaculture is heavily dominated by the Asia–Pacific region, which accounts for 89% of production in terms of quantity and 79% in terms of value, and is growing at more than 5% a year. This is mainly because of China, which accounts for 62% of quantity and 51% of value. Aquaculture production bordering the Atlantic is a minor component of global totals: Europe (3.6%), South America (2.2%), North America (1.5%) and Africa (1.4%). In the european union (EU), aquaculture currently provides 25% of fish protein [COM 12] and more than 90% of aquaculture businesses in the EU are small and medium enterprises (SMEs), providing around 80,000 jobs [COM 13].

8.2.5. *Offshore wind and ocean energy*

Offshore winds tend to blow harder and more uniformly than on land, providing higher potential for electricity generation – they are smoother and steadier compared to land-based wind energy. Globally, total installed offshore wind capacity was 3,117.6 megawatts (MW) in 2010, with 1,161.7 MW added in that year alone. The growth rate of 59% in 2010 was far above the growth rate of the wind sector overall. The North Atlantic has the potential to generate considerable renewable energy from offshore wind, especially during the winter. As of 2010, offshore wind farms had been installed by 12 countries, 10 of which were in Europe. A total of 10 gigawatts (GW) of capacity had been installed, led by the UK, Denmark, the Netherlands and Sweden. The EU has a target of 40 GW of offshore wind power capacity by 2020 and 150 GW by 2030 (source European Wind Energy Association: EWEA).

Figure 8.6. *Offshore wind mill: project Winflo (source: Winflo) in [PAI 14]*

The world's ocean waves, currents and tides are estimated to contain more than 5,000 times the current global energy demand, with estimates that marine resources could feasibly provide 20,000 TWh (terawatts-hour) of electricity per year – more than the entire global generation capacity. A variety of mechanisms are being developed to convert ocean energy efficiently from these raw sources into electrical power, and several devices are being tested (see Chapter 4 of [PAI 14]), but the engineering challenges for technology to survive for long periods in the harsh marine environment present many challenges. The maturation of ocean power technologies depends upon deployment of substantial demonstration and commercial projects in near shore areas. In the Atlantic, some of the greatest potential and need for ocean energy is in the Northeast, and this is where the majority of the research and development is taking place. Currently, there are only a few hundred MW worth of projects installed around the world, mostly in European waters (519 MW in 2011 mostly from tidal power plants [PAI 14]).

8.2.6. *Marine, coastal and cruise tourism*

The number of cruise ship passengers has grown nearly twice as fast as world international tourist arrivals from 1998 to 2008. With about 14 million passengers in 2010, the industry is expected to grow at 8.5% per year over the next decade. According to Cruise Industry News, in 2014, the 296 ships in operation had a capacity estimated at 21.4 million passengers grossing 33.8 billion USD[6]. The 100 plus ships of the main international cruise industry association[7] account for about two-thirds of the world's cruise ships, comprising less than 5% of all passenger ships and only 0.2% of the world's trading fleet. About 70% of cruise destinations are in the Caribbean, Mediterranean, Western Mexico and the South Pacific. In 2001, the North American cruise industry contributed 20 billion USD to the US economy, a 2 billion USD increase over 2000. Within Europe, cruise tourism employs nearly 150,000 people and generates a direct turnover of €14.5 billion, with the European market growing rapidly. Still, about half of the world's cruise passengers depart from US ports for the Caribbean.

Overall, in the Caribbean, tourism provides over 18% of regional GDP (and more than 50% in several individual nations), approximately 16% of employment, and 25% of foreign exchange earnings. Total tourism demand

---

6 Cruise Industry News, 2015.
7 CLIA: Cruise Lines International Association.

in the Caribbean region is currently 40.3 billion USD and is expected to grow to 81.9 billion USD by 2014.

Figure 8.7. *Cruise ship and tourism (copyright Wolcott Henry – Marine Photobank)*

Tourism receipts directly account for more than 75% of total exports and indirectly contribute to the growth of other sectors including agriculture, construction and manufacturing. Capital investment in the industry is estimated at 7.4 billion USD, or 21.7% of total investment, generating one in seven jobs in the Caribbean. In Europe, the coast is the preferred holiday destination of 63% of European tourists, where maritime and coastal tourism is the largest single maritime economic activity, employing 2.35 million people, equivalent to 1.1% of total EU employment. Cross-border coordination, as part of a sea-basin strategy, can contribute to the development of high-value tourism areas.

8.3. Catalyzing international ocean business leadership and collaboration

The World Ocean Council (WOC) was established to address the ocean sustainability issues and opportunities critical to business. The UN Secretary-General's 2010 report on oceans and the law of the sea noted that there is a need to "create awareness and understanding among industry of the ecosystem approach, marine biodiversity and marine spatial planning; develop regional ocean business councils; and strengthen efforts to create a global cross-sectoral industry alliance to constructively engage in United Nations and other international processes relevant to oceans, through organizations such as the World Ocean Council".

The WOC harnesses the potential for global leadership and collaboration in ocean sustainability, science stewardship by responsible ocean companies that are well placed to develop and drive solutions. Many companies want to address marine environmental issues, differentiate themselves from poor performers, collaborate within and across sectors and engage other ocean stakeholders. There is now a structure and process for companies to work on complex, intertwined, international ocean sustainability issues.

The WOC's international, multi-sectoral structure and process for leadership companies from the ocean business community is uniquely positioned to serve as a portal for this business community to work with other clusters and research institutions and consortia. A multi-sectoral and multi-stakeholder approach can result in cost-savings (e.g. collaborative research to develop best practices in sustainability and find science-based solutions to shared issues) and reduce the risk of costly, unplanned and unnecessary restrictions to responsible business operations in the marine environment.

Protecting the seas to protect your business makes good business sense, e.g. through the economies of scale that can be achieved in joint research on shared problems. Identifying problems and developing solutions must be based on good science, credible risk assessment, performance monitoring and the best available technology and must be tackled at the scale at which the impacts are accumulating.

Companies with a long-term view of their ocean business are also looking to collaborate within and between industries on solutions through participation in the WOC. This not only applies to the companies that directly operate and use marine spaces or resources, but also to the wide range of industries linked to, or dependent on, those direct ocean users. This includes marine technology, mining, manufacturing and many other sectors. In fact, any company that transports its products by sea is part of the associated marine environmental impacts.

To address priorities, the WOC has created cross-sectoral industry working groups in the thematic program areas that have emerged: ocean policy and governance; ocean/marine planning, operational/technical issues, e.g. invasive species, marine debris, marine sound and marine mammal impacts; regional interests, e.g. the Arctic, the Mediterranean, and the Caribbean; adaptation of ports and coastal infrastructure to sea-level rise/extreme weather events and the "Smart Ocean-Smart Industries" program for improving data collection and sharing across industries.

8.4. Smart oceans–smart industries: industry leadership to build ocean knowledge

With the advent of the WOC, there is now an organization that is uniquely positioned to catalyze the role of business in addressing a range of priority ocean needs and opportunities. One of these priorities is developing a system to coordinate the expansion and improvement of data collection by ocean industries.

The WOC "Smart Ocean-Smart Industries" program has been launched following discussions with key national and international ocean and atmosphere observation programs. Leadership companies from a range of ocean industries have encouraged the WOC to develop this portal to scale up data collection from vessels and platforms and to coordinate with the scientific community.

A large-scale integrated multi-industry effort to advance the role of ships and platforms in collecting data must employ standardized procedures, technologies and instrumentation. Collaboration will facilitate the development of sensors and instrumentation appropriate for harsh marine conditions and rigors of routine operation on commercial vessels and platforms, and also ensure easy installation, removal and servicing. Overall, the "Smart Ocean-Smart Industries" program can create synergies and economies of scale for developing technology, operational practices and institutional arrangements, both within key sectors, such as shipping and offshore oil and gas, and across the wider range of ocean industries.

Within the framework of broad-scale needs and opportunities for improved data collection by industry, it will be very important to develop a phased approach. This will enable leadership companies to focus on specific, implementable activities that deliver short-term outputs, e.g. demonstrating the ability to form the partnership starting with one ship, to install and operate with instrumentation collecting and reporting basic oceanographic data and then scaling this up to more kinds of data and/or more vessels.

It is critical to learn from and build on the existing Ship of Opportunity programs[8] and to work with and through existing national and international organizations that collect, transmit, store and analyze oceanographic and atmospheric information. In particular, this includes coordinating with the relevant programs at the World Metrological Organization (WMO) and the Intergovernmental Oceanographic Commission (IOC), especially the Global Ocean Observing System (GOOS)[9].

The business values of the program include: improved information for ocean condition observations; nowcasts, forecasts and hindcasts; improved predictability of, and reduced risk from, extreme events that impact ships and platforms; improved weather information and resulting savings from ship routing, fuel efficiencies, etc.; reputational benefits from contributing to ocean positive efforts to document and monitor the marine environment; opportunities for educational and promotional outreach to stakeholders and the public; increased leverage and opportunities to shape ocean science and policy; participation in the development of emerging observational technologies; increased data on the physical and biological environment in which commercial activities are taking place; standardized data on environmental conditions and impacts, e.g. air and water emissions; data-driven input to corporate policies and practices; an increased and improved science basis for interaction with stakeholders on marine environmental issues.

---

8 Such experimental programs are currently being, or have already been, developed. We could cite, for example, the GEPECO (*géochimie du phytoplancton et couleur de l'océan* (geochemistry of phytoplancton and color of the ocean)) project developed from 1999 to 2002 and which covered the North Atlantic, the Gulf of Mexico, the equatorial Pacific, Polynesia, and the Tasman Sea. The objective was to collect at the surface, every 4 hours and using a commercial traveling ship, information on the relationships between physical forcing, phytoplanctonic populations and the color of the ocean. The GEPECO project was one of the operations of a larger program called PROOF (*Processus biogéochimiques dans l'océan et flux* (Biogeochemical processes in the ocean and flows)) by the CNRS, IFREMER, IRD, CNES and TOTAL.

9 GOOS – Global Observing Ocean System – is a global variable that integrates marine variables, including: information from ships, fired or drifting buoys and subsurface floating devices, tide measurements, position and strength of currents, wave height, toxic blooms, estimates of the vulnerability of fisheries and aquaculture resources, etc..

The program's benefits to science and governments include: the ability to collect oceanic and atmospheric data on a significantly expanded spatial and temporal scale; the collection of data over longer time periods and/or along repeated routes; the observation of ocean and atmospheric conditions in ways and places impossible to reach by other means; the opportunity to fill major gaps in data and understanding; a highly cost-effective means of data collection; increasing the global scope, scale and perspective of ocean data and understanding; improving and expanding the partnership and common ground between science, government and industry.

A comprehensive system of oceanic and atmospheric observations and monitoring will also provide input to international conventions and treaties, including: United Nations Convention on the Law of the Sea (UNCLOS), United Nations Framework Convention on Climate Change (UNFCCC), Convention on Biological Diversity (CBD), International Maritime Organization (IMO) and Marine Pollution treaties (MARPOL).

The WOC "Smart Ocean-Smart Industries" program is being developed as a bold new initiative that will link the commitment of leadership ocean companies to improving ocean science and health with the scientific community that collects oceanic and atmospheric data to provide a better understanding of the ocean and climate. The program's vision is for leadership companies from a range of ocean industries to collaborate with the scientific community in the systematic, regular, sustained and integrated collection and reporting of standardized oceanographic and atmospheric data for input to scientific programs that improve the safety and sustainability of commercial activities at sea and contribute to maintaining and improving ocean health.

The program will expand the number of vessels and platforms used to collect standardized ocean, weather and climate data, improve the coordination and efficiency of data sharing and input to national/international systems and build on ships and platforms of opportunity programs. At the present time, the WOC is moving forward on this initiative and defining the next steps, such as the value proposition/rationale for industry and science, an inventory of existing ships/platforms of opportunity programs, the menu of options for voluntary observations, interface requirements for platforms/payload, the principles, practice and platform for industry data sharing and access, and regional pilot projects.

8.5. Ocean industry leadership and collaboration for a sustainable ocean future

The global ocean hosts increasingly varied economic activities, meaning that the role of industry is a key to ensuring ocean health. The private sector not only needs to ensure access and social license, but also to reduce its risks by implementing solutions to sustainable development and environmental challenges. The business value for the ocean business community coming from collaboration on sustainability, stewardship and science is a compelling factor.

The WOC, the international multi-industry leadership alliance of ocean companies, is a leadership opportunity for responsible ocean companies to address risks and opportunities and, most importantly, a powerful tool in ensuring good governance for sustainable marine development.

The growing ranks of WOC member companies are finding direct business benefits in the synergies and economies of scale in collaborating with like-minded peers in other companies on these shared ocean industry challenges. As a result, an increasing number and range of ocean companies from around the world are joining the WOC to advance industry leadership and collaboration in *Corporate Ocean Responsibility*.

8.6. Bibliography

[CNU 12] CNUCED, Etudes sur les transports maritimes en 2011, Rapport du CNUCED, Nations unies, 2012.

[COM 12] EUROPEAN COMMISSION, Blue Growth opportunities for marine and maritime sustainable growth, COM (2012), 0494 final, 2012.

[COM 13] EUROPEAN COMMISSION, Orientations stratégiques pour le développement durable de l'aquaculture dans l'Union européenne, COM (2013), 229 final, 2013.

[FAO 12] FAO, The State of the World Fisheries and Aquaculture 2012, FAO Rome, available at www.fao.org/docrep/016/i2727e/i2727e00.htm, 2012.

[FAO 14] FAO, The State of the World Fisheries and Aquaculture 2014. FAO Rome, available at www.fao.org/3/a-i3720e/index.html.

[GRO 14] GROS P., PROUZET P., "The impact of global change on the dynamics of marine living resources", in MONACO A., PROUZET P. (eds), *Ecosystem Sustainability and Global Change*, ISTE, London and John Wiley & Sons, New York, 2015.

[LAG 99] LAGRANGE X., GODLEWSKI P., TABBANE S., *Réseaux GSM-DCS,* 4th ed., Hermes, 1999.

[MON 14] MONACO A., PROUZET P., *Ecosystem Sustainability and Global Change*, ISTE, London and John Wiley & Sons, New York, 2014.

[MON 15] MONACO A., PROUZET P., *Development of Marine Resources*, ISTE, London and John Wiley & Sons, New York, 2015.

[OEC 14] OEC – Food and Agricultural Organization of the United Nations, *OECD FAO Agricultural Outlook 2014*, OECD Publishing, Paris, 2014.

[PAI 14] PAILLARD M., MULTON B., BOEUF M., "Marine renewable energies", In MONACO A., PROUZET P. (eds), *Development of Marine Resources,* ISTE, London and John Wiley & Sons, New York , 2015.

[PEL 98] PELISSIER C., *Unix: Utilisation, Administration, Réseau Internet*, 3rd ed., Hermes, Paris, 1998.

List of Authors

Frédérique ALBAN
AMURE
University of Western Brittany
Brest
France

Nicolas BOILLET
AMURE
University of Western Brittany
Brest
France

Jean BONCOEUR
AMURE
University of Western Brittany
Brest
France

Annie CUDENNEC
AMURE
University of Western Brittany
Brest
France

Olivier CURTIL
AMURE
University of Western Brittany
Brest
France

Cécile DE CET BERTIN
AMURE
University of Western Brittany
Brest
France

Florence GALLETTI
IRD–MARBEC
Sète
France

Gaëlle GUEGUEN-HALLOUET
AMURE
University of Western Brittany
Brest
France

Yves HENOCQUE
IFREMER
Issy-les-Moulineaux
France

Paul HOLTHUS
WOC
Hawaii
USA

Bernard KALAORA
LITTOCEAN–LAIOS/EHESS
Paris
France

Véronique LABROT
AMURE
University of Western Brittany
Brest
France

André MARIOTTI
UPMC–IUF
Paris
France

Jean-Baptiste MARRE
AMURE
University of Western Brittany
Brest
France

André MONACO
CNRS–Cefrem–UPVD
Perpignan
France

Arnaud MONTAS
AMURE
University of Western Brittany
Brest
France

Jean-Charles POMEROL
UPMC–INSIS/CNRS
Paris
France

Patrick PROUZET
IFREMER–DS
Issy-les-Moulineaux
France

Index

Other titles from

in

Earth Systems – Environmental Engineering

Lightning Source UK Ltd.
Milton Keynes UK
UKOW06n1928081215

264335UK00011B/88/P